Fiborn Quarry
Then and Now

Fiborn Quarry
Then and Now

Mark Whitney

Michigan Karst Conservancy
Ann Arbor, Michigan

Library of Congress control number: 2014937747

ISBN 978-0-9663547-9-9

First Edition August, 2014

Produced by the Michigan Karst Conservancy
Fiborn History Project

Cover Photographs by Tim Deady: The quarry works in 2012.

Book Design by
Greyhound Press
Bloomington, Indiana

GREYHOUND PRESS

Published by
The Michigan Karst Conservancy
2805 Gladstone Avenue
Ann Arbor, Michigan 48104-6432
USA

Forward

This book is the result of 25 years of gathering information, interviews, and photographs by several people who were interested in recording the history of the Michigan Karst Conservancy's first and, arguably, most important project. The data gathering was begun years ago by Mike Warner and Stanley Bell. Finally a member appeared who was able to digest all this information and produce an interesting book. Mark Whitney's effort is here before you.

The first effort at recording the history of the area is a book with the title *Tales & Trails of Tro–La–Oz–Ken* Published in 1976 by the Trout Lake Women's Club as a contribution to the bicentennial celebration of the United States. That 158-page book has only two pages referring to Fiborn Quarry. It is filled with stories of the history of the Trout Lake area preceding the opening of the quarry and is intended as a record of the families who settled and developed the land. It can be considered a record of the history preceding that recorded in this book.

The story of Fiborn Quarry is an exciting example of development of an industry in the wilderness. A whole town and industry developed in the north woods of Michigan's Upper Peninsula to fulfill the dreams and foresight of two creative individuals.

Tom Rea
Property Manager, Fiborn Karst Preserve

Contents

Preface

From the winter of 1905 until the early, wintry days of 1936, a town grew up around a "small hive of industry" atop thinly wooded ridges of limestone west of Trout Lake in Michigan's Upper Peninsula. Tiny even by U.P. standards, the Village of Fiborn Quarry (spawned by the business of Fiborn Quarry) was initially populated by dozens of single men living in a boarding house and bunkhouses, about a half-dozen women hired to cook and clean, and just a small handful of families. Over time, families became the resident majority and the numbers of single, itinerant workers bouncing between rock-crushing and timber-cutting fell to less than a dozen. People worked hard, often growing, canning, hunting, and trapping much of their food. They played baseball and held square dances for entertainment and tried to keep their financial heads above water despite sometimes falling behind at the company store. Children went to a one-room schoolhouse, played in nearby woods and streams, and often worked hard like the adults. A few babies were born and a few men died in gruesome accidents.

After the Michigan Karst Conservancy (MKC) bought most of the Fiborn Quarry property and established the Fiborn Karst Preserve, members formed the Fiborn History Project to collect and preserve stories of the lives lived in the tiny village whose story follows much the same arc of late 19th and early 20th century boom towns across the U.P. Partly, the history project sprang from the MKC's primary mission of preserving the area's karst features: Documenting its history was part of this stewardship. But members also became personally interested in the people and stories of Fiborn, in large part because in the late 1980s, quite a few former residents were still alive and living in the eastern U.P.

MKC life member Mike Warner heads the long list of contributors to this history of Fiborn Quarry. Warner scoured U.P. newspaper archives on microfilm, the extensive papers of onetime Michigan Governor Chase Salmon Osborn (at the University of Michigan's Bentley Historical Library), and along with fellow MKC member Stanley Bell, spent most of several summers interviewing former Fiborn residents, the descendants of former residents, area residents who knew about the quarry, and others with indirect knowledge, such as men who'd worked at other quarries or in lumber camps near Fiborn. With ties to the Beaudoin family of Trout Lake (descended from Fiborn Quarry employee Oliver Beaudoin), Stan often made the connections and introductions, while Warner conducted and recorded nearly 24 hours of interviews and transcribed many of them.

Warner and Bell also travelled to Sault Ste. Marie, Ontario, and Algoma Steel Corporation (which owned the quarry from 1910 until its closing), where they found and studied what business records the company still had. Warner also studied state geologists' reports on the Fiborn area at the state of Michigan Archives in Lansing.

Other MKC members conducted or helped conduct interviews, including Barb Patrie and Aubrey Golden. Warner and Golden also travelled to New Jersey and Bruce Mines, Ontario, following the trail of the quarry's first superintendent.

The project produced articles and the display at the preserve's Emma Kalnbach Pavilion (named after the village's last school teacher, who contributed interviews and photos to the project), but stalled for several years.

Joining the project in 2011, MKC life member Mark Whitney retraced and expanded on Warner's research in the Chase Osborn papers, as well as newspaper archives in Sault Ste. Marie, St. Ignace, and Newberry. He also researched census rolls from 1900 through 1930 for Mackinac County, Sault Ste. Marie, and Marquette; studied histories of Michigan and the Upper Peninsula, Algoma Steel, and the Duluth South Shore & Atlantic railroad company (each cited extensively in the endnotes); and discovered a cache of business records at the University of Western Ontario and mentions of Fiborn in mining and industrial journals of the time (also cited in the endnotes). He reviewed Warner's taped interviews, studied transcriptions of several interviews and transcribed three interviews; studied cemetery records; then finally organized and wrote the historical narrative. With the help of MKC member Betty Smith, Mark was also able to re-obtain a valuable collection of photos of the quarry taken in 1924, and scan them at high resolution for publication.

MKC Trustee Ruth Vail edited the manuscript and offered valuable suggestions for handling the large, but sketchy amount of information that sometimes threatened to bog down the narrative in bits of unconnected detail.

Rane Curl researched and wrote a chapter on the geological setting of Fiborn Quarry, explaining why the limestone outcrops on which the company and village were literally built held enough value to spark a small boom-town for a short time.

MKC life member, Trustee, and Property Manager of the Fiborn Karst Preserve, Tom Rea—proprietor of Greyhound Press in Bloomington, Indiana—copy edited and typeset the material and arranged for its publication as this book.

Collected mostly from scattered pieces of historical record, with only the last few years of the story livened with personal memories, this history of Fiborn is far from complete, but hopefully draws together threads and patches from many stories into an imperfect but warm, comfortable, and even picturesque quilt, like you might find on a bed in a worker's small, wooden home along the corduroy road into Fiborn Quarry around 1929.

Then: A History of Fiborn Quarry

Surveyors and Speculators

Henry B. Brevoort Jr. was a United States deputy surveyor trudging through forests and swamps to cut, measure, and mark section lines across Michigan's eastern Upper Peninsula in the spring and summer of 1845. While laying the south line of section 16 through forest in the 44th township north of Michigan's survey baseline and seventh west of its prime meridian, Brevoort traveled over a natural "bridge" of limestone separating a pair of sinkholes, one of which led to an underground passage. "Cave. Entrance 2 feet by 3," he wrote in his field notes. Caves are rare in Michigan, so such a sight would be worth noting, but it was also Brevoort's job. Surveyors' duties included keeping detailed inventories of natural features such as streams, ravines, hills, mineral and timber resources, and soil types.[1]

If he investigated the cave, Brevoort would have noticed it played a role in draining water from the large swamp running to the south and west into the river lying east of the rocky outcrops. He also may have noticed some particularly pure limestone, the kind eagerly sought by manufacturers of steel and calcium carbide, and concluded that someone could run a profitable quarry if there were nearby transportation, such as a railroad line. The land was rich in timber, which could be used to build infrastructure and sold to generate cash, and had flowing water and good drainage.

The railroad came some 35 years after Brevoort found the cave entrance. A line from Marquette to St. Ignace was finished in 1881 with passenger service beginning in December, and a branch added in 1887 linking it to Sault Ste. Marie.[2] The Detroit Mackinac & Marquette railroad company, which built the line after obtaining a federal land grant, sold land and tried to entice settlers along the route, but the region's soils, swamps, and short growing season discouraged farming, and land sales for settlement were slow to develop. The DM&M struggled financially and in 1886 was reorganized then merged with three other regional lines

9

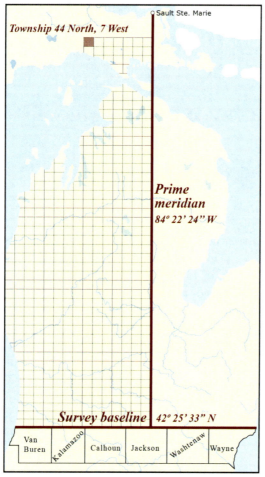

Township 44 North, 7 West

Sault Ste. Marie

*Prime
meridian
84° 22' 24" W*

Survey baseline | *42° 25' 33" N*

Van Buren | Kalamazoo | Calhoun | Jackson | Washtenaw | Wayne

*Surveyors laid a vast grid of townships and sections across Michigan's wilderness
from the 1830s until 1851. The area that became Fiborn Quarry was
surveyed by Henry Brevoort in the summer of 1845.*

to form the $22 million Duluth, South Shore & Atlantic, which for years profited by hauling logs and minerals during the Upper Peninsula's lumber and mining booms. The line connected Duluth, Marquette, St. Ignace, and Sault Ste. Marie, creating a steady flow of commerce between them and Michigan's Lower Peninsula by ferry across the Mackinac Straits.[3]

Timber was the wilderness area's most valuable resource, so lumbering companies and individual speculators made up many of the early buyers. One of those individual speculators was a Charles D. Rood of Springfield, Massachusets, who in 1893 bought 120 acres in Section 16, T44N, R7W,

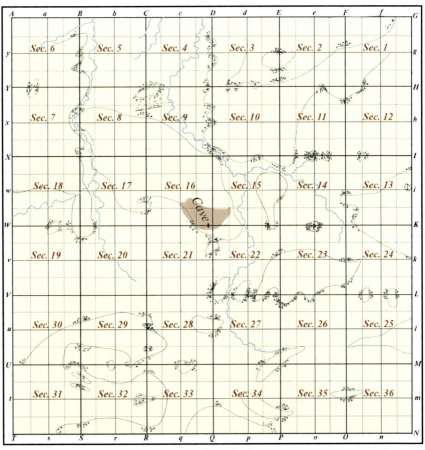

Township 44N, Range 7W, later part of Hendricks Township, with the general area of the quarry that would grow up around the caves discovered by Brevoort.

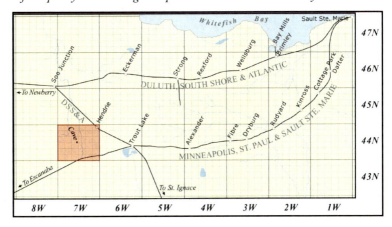

Eastern U.P. with 1881 rail lines and stations

11

including the southeast corner of section 16 where Brevoort marked the cave entrance. Rood, who appears to have never visited the area, probably employed a "timber cruiser" to scout it, a common practice during the lumbering era.

The Iron Hunter and the Railroad Man

Hundreds of millions of years' worth of geological workings left a valuable natural resource—outcrops of highly pure limestone—jutting up into the scrubby pine land Rood purchased. Extracting it would take considerable effort, expense, and ingenuity. The job required a prospector's skills and stamina, a lawyer's eye for legal detail, a salesman's bravado, and a CEO's power to marshal money and manpower. Beginning in late 1898, a firebrand future governor of Michigan provided the first three, and a distinguished railroad leader delivered the fourth.

Chase Osborn   *William Fitch*
Chase Osborn collection, Bentley Historical Library, University of Michigan

Chase Salmon Osborn and William Foresman Fitch proved to be effective partners: Osborn—a 39-year-old state game warden, newspaper publisher, explorer, and speculator—patched together 740 contiguous acres in the sometimes-foggy legal wake of a land rush and actively promoted it; Fitch, a 59-year-old veteran railroad leader, was the operational force when it became needed, building the all-important and expensive 3- to 4-mile railroad spur, recruiting and dealing with

12

contractors, and overseeing the first quarry work.

Osborn, an Indiana native, came to the U.P. in the late 1880s by way of Wisconsin. Starting out as a reporter, he later owned or co-owned several newspapers including the *Sault News* in Sault Ste. Marie. He also was a rising star in Michigan Republican politics, appointed state game warden in 1895 and state railroad commissioner in 1899. As game warden, he was based in "the Soo," but as railroad commissioner he commuted to Lansing, keeping his home and family in Sault Ste. Marie[4] and selling his interest in the *Sault News* in 1901. An avid outdoorsman, he hunted about 10 miles northwest of the cave at his Deerfoot Lodge.

Fitch was president and general manager of the Duluth, South Shore & Atlantic, based at company headquarters in Marquette, where he lived with his family.[5] Like Osborn, he had a range of outside business interests.

The two were friends by the mid-1890s. Osborn sought and received rail-travel favors from Fitch, and often replied with gifts and letters professing a deep, affectionate friendship.[6] Osborn became Michigan's railroad commissioner in 1899 and battled mightily with companies across the state over safety and service (he was very much in the mold of reform-minded politicians such as Theodore Roosevelt, to whom he was often compared), but this never affected his relationship with Fitch, judging from their correspondence.[7]

Osborn, went on annual prospecting expeditions in the Upper Peninsula and Canada, mostly in pursuit of iron ore. (He titled his 1919 autobiography *The Iron Hunter*.) He speculated in limestone during the late 1890s as local industrial growth drove up demand. Fitch also was looking for limestone that could be mined. Along with steel makers that used crushed limestone to remove impurities from molten iron; makers of calcium carbide, widely used in acetylene lighting; and producers of the fertilizer calcium cyanamide, were among the potential customers for high-quality limestone deposits. In 1898, Union Carbide Company absorbed the Lake Superior Carbide Company in Sault Ste. Marie, Michigan, and planned to build a plant there to manufacture calcium carbide.[8] (The Chapter "Geological Setting" discusses this in detail.)

Osborn explored what would become Fiborn Quarry in late 1898, with Charles B. Smith, superintendent of a lumber camp owned by Osborn's friend and fellow Sault Ste. Marie businessman Frank Perry. Osborn traveled by train December 20 from the Soo to a junction near Perry's camp west of Trout Lake and spent the night. The next day, Osborn and Smith rode to Lewis Station, south of the cave, then hiked. Osborn, a relentless diarist, wrote that he rose at 4:30 A.M. and took the train "with Mr. Smith to Sec. 16-44-7. Big Cave. Walked 4½ mi. and carried [limestone] rock. Got home 7:45. Worked in office. Hard day. Bed 1:45."

The Iron Hunter must have been impressed with the limestone rock he and Smith carried from the "Big Cave." He immediately sought the identities of landowners and made offers. Much of the land was held by the state over delinquent taxes, and Osborn acquired rights to it by paying the taxes owed, eventually buying it for $4 per acre.

The Chase is On

Osborn wrote to Michigan Land Commissioner William A. French in Lansing on December 22, asking "what portions, if any," of 16-44N-7W were owned by the state, their sale price, and who may have bought other parts from the state. Osborn also wrote to the Mackinac County Register of Deeds in St. Ignace two days later asking for names and addresses of landowners in Section 16.

He discovered Charles Rood immediately and wrote to him on Christmas Eve, asking if his land was for sale. "I take it for granted that you know the pine has been cut off of your land and that it now only rates as second rate hardwood land, in addition to being very stony," Osborn wrote, pressing for a low price. He also claimed he was "thinking of building a hunting lodge."

Osborn wrote Fitch three days later, noting that the latter had been investigating limestone prospects in nearby Luce County. "I hope you will not go into it very far until I have seen you, because I think I can tell you where there is stone much better than that and nearer to the Soo and nearer to the railroad."

Osborn wired $320 to the land commission office on December 29 to buy two 40-acre parcels, then sent a check for $960 to buy six others on January 24, 1899.

Rood replied January 11, and in what would become a familiar pattern, begged Osborn to "excuse the delay which has occurred in replying to your favor of the 24th." He asked how much Osborn was willing to pay, and if the hunting lodge was to be a commercial venture (which naturally would drive up the price). Osborn replied two days later, offering "$1.25 an acre for your three forties" in cash and insisting he planned "simply an outing camp in the wild woods."

Fitch and Osborn formed a partnership to acquire the land and promote a quarry. Osborn wrote in his diary that on January 23 he "sold Mr. Fitch ½ int. in 8 forties, Sec. 16-44-7 for $1280—Limestone, etc." Osborn sent Fitch a letter the next day acknowledging the check, and "as a matter of memorandum," laid out their basic agreement: Osborn would "secure title direct from the state of Michigan" and deed Fitch "one half of the same." They

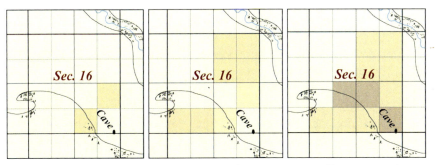

Left map: Osborn wired $320 to Michigan Land Commissioner William A. French on Dec. 29, 1899, to buy the northeast and southwest quarters of the southeast quarter of section 16.

Center map: Osborn mailed $960 on Jan. 24, 1900, to buy the northeast quarter and the south half of the southwest quarter. He sold Fitch a half-interest in the land bought to date for $1,280.

Right map: Osborn had a hard time buying clear title to the three forties owned by Charles Rood of Massachusetts. Throughout 1899, He pursued Rood, then Newberry-area lumberman Isaac Wilman to buy the land.

were to equally divide the tax bills and any money made if they sold the land.

Rood rejected Osborn's offer of $1.25 an acre, and Osborn wrote back January 24, expressing his regret. Rood, who was stalling because he couldn't show clear title to the land and was scrambling to obtain it, suggested they meet in person. Osborn, who was planning a trip to Washington, D.C., and New York City in March, said he would consider it. Osborn also wrote to Fitch in early February, noting, "It may be that I will have to run down to Massachusetts" and asking if Fitch could obtain passes from the various railroads he would use. Rood devoted a couple of letters to little but small talk about hunting, which Osborn answered with his usual mix of friendliness and frustration, offering a "standing invitation to come to my hunting camp" and possibly sending along a package of venison, before pleading with Rood to hurry up and set a price.

Osborn left Sault Ste. Marie with his family on February 27, arriving in Washington on March 4, according to his diary. After several days of meetings with government officials and political figures, he set out March 10 for Massachusetts and arrived at Springfield the following afternoon to find Rood had dodged him. "Only phoned Rood," Osborn wrote. "He was at Middletown," Still, it seemed the trip wasn't in vain. "He said he would let me have the land," he added.

Though Osborn probably arrived home satisfied he'd finally put the key piece of his puzzle in place, it still lay just outside his reach. Rood continued to stall, and two weeks after their first conversation, Osborn phoned again while on another trip to New York. "Called C.D. Rood [in] Springfield, long distance, about the land," he wrote in his diary, offering no hint about how the conversation went. He wrote to Rood for the final time nearly a month later, and apparently received no reply.

Osborn put his right-hand man at the *Sault News* on the case. In May, his secretary, Mary Frances Hadrich, wrote to Managing Editor Ethan W. Kibby, asking that he investigate ownership of the three forties. Rood, meanwhile, had St. Ignace attorney and longtime Mackinac County Prosecutor Henry Hoffman trying to regain the deeds from a stubborn lumber man living deep in the woods southwest of Newberry, to whom Rood had relinquished them.

Osborn made one more trip to Springfield on July 18 while again visiting New York. He wrote in his diary that he took a train to Springfield, arrived at nearly 3:30 P.M., saw Rood, and left for New York at 4:30 P.M. Rood may have finally admitted during their brief meeting that he didn't have clear title and was trying to get it back. In a letter months later to Fitch recapping his efforts, Osborn said: "I finally learned that he had quit-claimed them and was trying to buy them back to sell to us."

After Rood bought the land in 1893, he contracted "with parties who wanted the timber," Kibby told Osborn in a letter after several months of research. Rood's contract "expired before the contractors took off the timber, and in 1896 Rood quit-claimed the tract to (Isaac) Wilman." After Wilman cut and sold timber, "the party to whom the first contract was given stepped in and legal complications arose about ownership of the timber." Kibby said Hoffman offered $150 to Wilman on Rood's behalf, but Wilman did nothing.

Osborn had in fact contacted Hoffman well before he assigned Kibby to investigate the Rood/Wilman land, and by September, 1899, had the St. Ignace attorney negotiating with Wilman. Hoffman explained the situation much as Kibby later reported. "I made an agreement with him to pay him back $150 and he to quit-claim the land back," Hoffman wrote September 21. This agreement, however, wasn't in writing except in the attorney's personal letter book. "I sent Mr. Wilman a deed at that time to execute and he excused the execution because it was not convenient" and the matter had languished since. "He is a stubborn fellow and if he thinks there is anything in this land, he will impose harsh terms." About two weeks later, Hoffman reported: "I have written Mr. Wilman a very strong letter ... and if that don't fetch him, I can do nothing." The letter

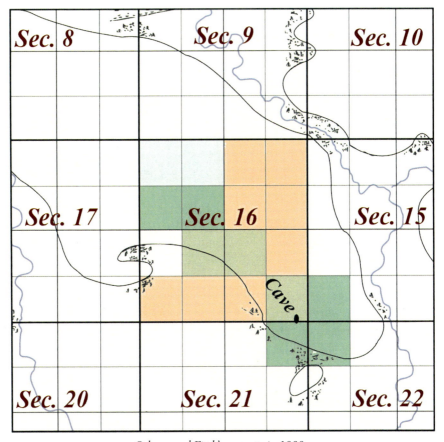

Osborn and Fitch's property in 1900
Paid $1,280 ($4 an acre) to state of Michigan.
Paid $175 ($1.45 an acre) to Isaac Wilman for quit-claim deed, and $52.67 in taxes and interest for a total of $227.67. Previously owned by Charles Rood.
Paid $411.64 ($2.57 an acre) to Hall & Munson lumber company and $63.36 in taxes and interest for a total of $475. Originally gained tax deed.
Paid state $32.52 in back taxes and $37.50 in recording and legal fees to obtain tax deed. Notes about the land in Osborn's papers say the tax deed was issued in error, and he released the 80-acre parcel back to the state in February, 1904.

scolded Wilman for talking about friendship, but not acting as a friend.

Hoffman's letter failed to fetch the stubborn fellow, so Osborn went to see him. On November 1, after spending the night at a Newberry hotel, Osborn "left 8:00 A.M. for 18-mile drive [train ride] to Wilman Mill near Whitefish Lake," he wrote in his diary. "Saw Wilman and got option on limestone land. Venison for dinner. Left 5:20 on DSS&A for home." In his letter to Fitch much later recapping the events, Osborn said he found Wilman "in the deep woods 30 miles southwest of Newberry. I drove out there in the roughest sort of weather and found that he had given an option to some lumbermen and mill owners on the Soo Line. I procured waiver of their option, and after three trips succeeded in getting a quitclaim from Wilman."

Osborn also enlisted the services of Newberry attorney and Luce County Prosecutor Furman E. Dutcher, who had represented Osborn previously in legal matters there. After one failed attempt, Dutcher reported obtaining Wilman and his wife's signatures on the deeds in a November 14 letter, noting "I had quite a jig with him over the $200 but finally gave him $175."

Even then, it wasn't over, thanks to an error in the legal paperwork. Kibby discovered that the quitclaim Wilman obtained from Rood mistakenly described his holdings as including the northwest quarter of section 16, rather than the northwest quarter of the southeast quarter of the section. Osborn told Kibby in a December letter that the Mackinac County deputy register of deeds in St. Ignace told him most likely an error had been made there in copying the deed. Dutcher had to get the Wilmans' signatures on corrected deeds, which he was finally able to do in January, 1900.

Meanwhile, Osborn received confirmation that the limestone he'd found would be valuable to carbide and steel manufacturers. Lake Superior Power Company analyzed two samples he had sent, and reported in a November 21 memo that one was only 83 percent calcium carbonate, not nearly pure enough for industrial purposes, but the second was a more-than-satisfactory 98.55 percent pure $CaCO_3$. (A second analysis was performed later, upgrading the first sample to a satisfactory 96.56 percent pure, and maintaining the second sample as slightly more than 98 percent pure.)

After waiting for deeds to be copied and corrected, and with enough land put together to begin promoting a quarry, Osborn on May 21, 1900, sent Fitch a statement of "amounts paid for limestone lands in Mackinac County" totaling $772.69, and seeking Fitch's share of $386.34. Osborn listed the parcels as purchased from the state, Wilman, and the Bay Mills-based Hall & Munson lumber company.

Amid his work as railroad commissioner; his timber land dealings and iron prospecting (he discovered a tract near Sudbury, Ontario, that would eventually profit him handsomely); and the teeth-pulling needed to obtain the Rood/Wilman land, Osborn ran for governor in 1900. He announced his candidacy in a wave of January letters to Republican figures and friends across Michigan and was a frequent speaker at GOP

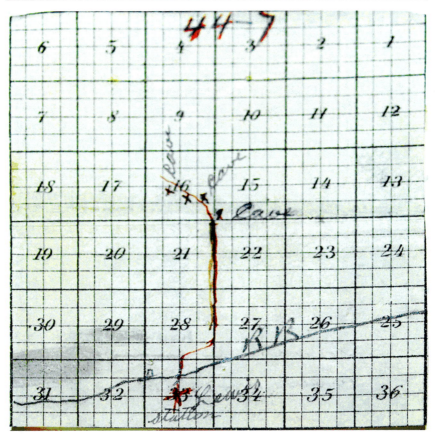

Charles B. Smith's map showing the route he and Chase Osborn took from Lewis (later Caffey) station to the southeast corner of section 16 and several cave entrances.

meetings around the state, but did little campaigning besides making news as railroad commissioner.[9] At the convention in Grand Rapids, Saginaw lumberman and banker Aaron Bliss was nominated (and later elected), and Osborn made news by accusing other candidates' supporters of buying votes.[10]

The Big Cave

Along with the prospect of selling the land for a limestone quarry, Osborn also took an interest in its caves. "I have made an interesting discovery of a very considerable cave which I have not had time yet to explore and which is located about ten or twelve miles from our hunting camp," he wrote to Deerfoot partner and Detroit industrialist R.J. Cram shortly after his trip with Charles Smith. "I went underground in it fully 300 feet I should judge and was compelled to back out on account of the insufficient torch light."

In January, 1900, while waiting to obtain title to the Rood/Wilman land, Osborn wrote to Smith and to a man named L. Leslie at Lewis Station, seeking information such as precise locations and detailed descriptions. Smith sent a map showing their route from Lewis Station to the area and marking several cave entrances.

"The mark on the extreme S.E. of S.E. is where we looked into cave and heard the running water," Smith wrote on January 30, in reply to a letter from Osborn.

"There [are] three that I know of," Leslie wrote February 1, "I have been through two of them. The largest one has a room 10 x 12 ft. and is 100 feet long and the other one [in] some places is 12 ft. high and tapers down from 8 ft. to 2 ft on the bottom. You can travel in this one about 1/4 of a mile the entire length of the other one with the large room is about 40 rods."

At State Railroad Commissioner Chase S. Osborn's invitation, State Geologist Alfred C. Lane visited the caves August 3, 1901, mapping the area's surface and cave features. "On land belonging to Chase S. Osborn, is an interesting group of caves in this limestone," Lane reported. "Not far from the southeast corner of the section the trail passes between some small sinks, which are the entrance to several hundred feet of cave, low and flat, in general not over 2 or 3 feet high, but with a channel 6 to 8 feet deep cut in beautiful meanders which slightly increase the size of the loop as they cut down, and are barely wide enough for a man to walk in them. The stream finally falls by a cascade into a larger and picturesque sink hole about 30 feet deep, in which no outlet could be found. It was probably concealed by broken rock." Lane also noted a pair of sinks and a river cave a few hundred yards northwest of Osborn's Big Cave. (Lane, in a handwritten note to a fellow geologist, said: "Had a great time in the Niagara Limestone caves, up to my armpits in 57-degree water, several hundred feet of galleries.")

"This limestone is said to be very pure and suitable for the manufacture of calcium carbide," Lane's report concluded. "There must be a considerable area, with little stripping and easily drained." Osborn was probably hoping he'd say that. Lane's trip was covered by the *Lansing Journal*, which published a story headlined "Caves in Michigan, Discovered by Chase Osborn in the Upper Peninsula," on August 10. "At the time of Dr. Lane's visit only a few hundred feet could be traversed before the waters rose breast high and steadily grew deeper," the story said. "The chamber could be seen going in for some distance and would even make an interesting boating trip."

From Prospecting to Promoting

Osborn and Fitch didn't intend to develop a quarry, just sell the land once they put the necessary parcels together. "A number of inquiries have followed the visit of Dr. Lane to our limestone property," Osborn told Fitch in a letter about two months after Lane's visit.

Despite the interest generated by Lane's trip and Osborn's promotion, he and Fitch weren't able to sell the land, and endured a series of disappointments from 1901 to early 1904.

- **October, 1901:** Osborn offered the land to Philadelphia-based American Alkali Company for $12,000 in a letter to B.E.F. Rhodin, managing director of the company's Sault Ste. Marie operations. Rhodin didn't reply by Osborn's November 15 deadline.
- **June, 1902:** Osborn again offered the land to American Alkali, and again asked for $12,000 in a June 23 letter to company President Arthur K. Brown. The offer was good until the New Year. Though Osborn told Fitch in a June 30 letter "I think the trade will go through," Brown made no move to accept the offer.
- **April, 1903:** Fitch and Osborn drew up an agreement to lease the land to William Rosevear, who planned to quarry limestone and sell it to G.W. Mead of Chicago. Though they signed a contract on April 9, the deal fell through and Osborn wrote to Fitch in July: "I had expected our limestone property would have been sold ... However a deal has not been consummated yet. I am working on the matter and considering several propositions."
- **February, 1904:** Osborn promoted the land to Union Carbide in a letter to General Manager W.P. Martin. Martin promised to bring the matter to the company's board, and finally replied in April that "at present at least our Directors are not willing to authorize the purchase of your land."
- **April, 1904:** George Nicholson, head of White Marble Lime Company in Manistique, took an option to buy the property, but surrendered it in June.

Finally, in the summer of 1904, Osborn and Fitch received the inquiry that would make Fiborn Quarry happen, though not as they had originally planned. Osborn contacted Algoma Steel in Sault Ste. Marie, Ontario, and reported to Fitch in a June 25 letter that Francis H. Clergue, head of Algoma's parent company, Lake Superior Corporation,

21

expressed interest in the limestone, and wanted to know Fitch's terms for building a railroad spur to the property and rate for hauling the quarried stone.

"He says that about 300 tons a day will be used at the Sault, and that other demands ought to easily bring the shipments up to 500 tons a day," Osborn wrote. Fitch estimated the rail-spur cost at no less than $8,000 per mile, or $40,000 for 5 miles, which he said they couldn't afford to advance. Within a few days, Fitch and Osborn met, and Osborn wrote to Clergue on July 1, saying Fitch agreed to "furnish the rail and build the bridges etc., upon such a spur, if you would grade it and tie it, and then he will make a rate to the Soo of 40 cents a ton on 300 tons a day."

Clergue didn't immediately reply; an apologetic reply to Osborn from his brother Bertrand said Francis was consumed by other business. In fact, Francis had all but lost his grip on the company by then, holding a largely advisory board seat after being ousted as president in April following Lake Superior Corporation's bankruptcy reorganization in February.[11]

Algoma Steel Receiver C.D. Warren, who became president of the parent corporation after Clergue's ouster, wrote to Osborn on July 25 outlining his company's requirements: 28,000 to 30,000 tons of rock a year; a contract for one to three years; a minimum purity of 90.5 percent calcium carbonate; and the stone crushed so that all pieces pass through a 3-inch ring. (Steelmakers add crushed, "burned" limestone to molten iron, where it reacts with impurities such as sulphur and phosphorous, forming slag that floats to the surface and is skimmed away.)

Though they had initially hoped to sell the property and take a profit, Fitch and Osborn now found themselves nurturing a startup business by the late summer of 1904, arranging logistics, searching for operational expertise, and rounding up prospective customers.

Dynamite, Steam, and Sweat

*T*he *Evening News* of Sault Ste. Marie, Michigan, (successor of Osborn's former paper) publicly announced the launch of what would become Fiborn Quarry on September 24, 1904. "Within a few weeks time a new limestone quarry will be opened in the vicinity of what are known as the Osborn caves, several miles west of this city," the paper reported, speculating that new venture would "put considerable money into circulation, which will naturally find its way to this city. ... A track will be laid to the quarry and a score or more of men will be employed."

Fitch took on the task, and apparently the cost, of building the railroad spur, likely limiting the latter by using Duluth, South Shore & Atlantic (DSS&A) workers and material. He also took on the task of hiring someone to run the quarry. Fitch first contracted with a John J. Case, possibly of Sault Ste. Marie. Case was mentioned in *The Evening News* article as one of the quarry's "interested parties" but nothing is known about him outside of a few letters between Osborn and Fitch.

"Mr. Case was [in Marquette] yesterday and we had a memorandum of the lease drawn," Fitch wrote to Osborn three days after *The Evening News* story ran. Every six months, the partners were to be paid at least 5 cents in royalties for every ton of rock Case sold for at least 45 cents. He was to quarry at least 20,000 tons of rock annually after the first year, and 25,000 tons after the sixth. He was responsible for property taxes for the 20-year life of the lease, which he could terminate with 90 days' notice.

Presumably, Case went about preparing to develop a quarry. Apparently, he did little but think about it. By his own account later, he spent much of August and part of September travelling to Duluth; Houghton; Detroit; and Sault Ste. Marie, Ontario, in addition to Marquette to see Fitch. However, Osborn and Fitch saw nothing accomplished during a visit to the site while the rail spur was being built. In a letter, Osborn recalled later that they "found two or three men who had been sent on the ground by Mr. Case doing nothing and aimlessly awaiting orders which never came."

Case also balked at signing a contract. October passed with little

23

or no work done. On November 15, a DSS&A employee telegrammed Fitch, saying Case was headed east. "Says he does not see his way clear to sign the contract as presented by Mr. Osborn, but is anxious to carry the thing through to a finish. Will have men in there this week Thursday to put up quarters for the winter and go ahead with the work, but will have to be reimbursed for any outlay he makes on the property. ... He will be in touch."

Fitch, however, had a legal order written denying Case and any of his employees the right "to erect buildings, put up machinery, clear land, or do any other work" on the quarry property. They also were ordered to leave and not return without permission. A DSS&A employee reported to Fitch he read the order to Cases's two employees on November 16 and they "left for the Soo that night."[12]

Case, probably unaware of the events of the 16th, telegrammed Fitch two days later: "Cannot accept contract however will personally superintendent completing of buildings by time tracks are ready you must furnish actual cash expended. ANSWER." Fitch replied that same day: "As you do not intend to sign contract, we will not allow you or anyone else to erect buildings or set up crusher on lands mentioned in contract without first signing contract with us." Recapping his dealings with Case in a letter to Osborn, Fitch said he told his men on the site "to take the logs that Mr. Case's men have cut and build two houses, regular logging camps, 16 or 18 by 30 at least."

Case eventually turned up in Lima, Peru, and submitted a bill for more than $1,000, which Fitch and Osborn ignored.

The Quarry is a Quarry Now

As Osborn and Fitch severed ties with Case and his paltry crew in November, 1904, Osborn received a letter from John Conlin, whose brother William ran a quarry in Hendricks Township for Union Carbide, asking for the job of running the new quarry. Osborn forwarded the letter to Fitch.

In late December, Fitch and Conlin drew up a shopping list of tools and equipment which included black powder, fuses, and electrical gear for blasting, as well as crowbars, sledge hammers, shovels, and wheel barrows for breaking and hauling rock. Another list of tools and equipment used to construct two log buildings was obtained from a storekeeper in Marquette. In a December 22 letter, Fitch relayed the lists to Osborn and said he would borrow a "steam drill and all the fittings" from Lake Superior Iron Company. The next day, Fitch wrote that he had sent for Conlin and would give him credit at the

Soo Hardware Company for the equipment on the list. The day after that, Fitch outlined his plan to open the quarry in a letter to Osborn. "I will arrange to meet you and Mr. Conlin and his men at Soo Junction next Wednesday morning the 28th ... I will have the boarding train there ... and go right into the quarry." Conlin presumably bought the equipment, transported it on his boss's railroad, and assembled a crew of workers (Fitch sent him rail passes for 10 men).

Fitch invited Osborn to the quarry in a January 4 telegram, but Osborn was to be in Duluth that day closing a deal for timber land east of Newberry known then as the Sage Tract. Two days later, Fitch wrote to Osborn: "The steam drill was started yesterday at the quarry. I had Mr. Johnson of the Newberry Furnace there and he wants one car a day and would like it as soon as we commence to ship. Mr. Conlin said he would be ready by Monday or Tuesday. Mr. Johnson said he would crush the stone and that we need not bother crushing it with hammers."

Conlin's crew drilled and blasted free its first rock by January 11, 1905, when Fitch wrote to Osborn: "We have a face 50 feet long and 5 feet deep opened and at least 10 cars of stone down and ready to load." Fitch also mentioned he was leaving the next day for Texas and would return "in time to pay off and look after the selling of the limestone."

"Indeed the quarry is a quarry now," Osborn replied, noting that he was leaving soon on a trip abroad until May. The partners apparently were confident about leaving the operation in Conlin's hands for a while.

Fitch continued to seek a contractor who could build a large operation out of the small quarry, while Osborn spent a few months traveling in Asia and the Pacific. By March, Fitch was dealing with Samuel Benjamin Martin, an Ohio native and established quarry operator then based in Dayton who had been in the business since 1887 or 1888.[13]

Martin, who was likely 38 years old at the time of the deal (he was 43 when the 1910 census was taken), signed a contract drawn up by Fitch on March 21.[14] Martin agreed to install a plant that could crush at least 400 tons of limestone per day, and took on all responsibility for developing the quarry as well as the liability of running it.

Little record survives of Martin, especially any personal details. A 1917 article in the *Republican-News* called Martin "the acknowledged master in these parts of all the ins and outs of limestone quarrying," but little else was reported about him over the years other than his comings and goings on business. Even Fitch and Osborn's letters contained no details or impressions of the man they hired to build Fiborn Quarry

25

into a major mining operation.

Martin built a typical quarry for the time, drilling holes in which to drop dynamite and blast long layers of rock away from outcrops, using steam shovels and train cars to get the rock to a crusher, then sorting the crushed rock and loading it onto trains for shipment to customers. New machinery would come and old machinery would go, horses would yield to steam engines, but the process would remain the same for slightly more than three decades.

In May, 1905, Osborn and Fitch incorporated their limestone company. They met May 21, and chose the name Fiborn. As Osborn noted in his diary, "Fiborn = Fitch + Osborn." The name passed into geology in 1915 when State Geologist R.A. Smith named the long formation on which Fitch and Osborn's quarry (and several others) sat the "Fiborn Limestone."

The quarry again made the front page of *The Evening News*, which reported on May 25: "Village of Fiborn Quarry newest thing on the map." The "small hive of industry, which gives promise of rapid future growth" was producing 150 tons of quarried rock daily, according to the newspaper, with the goal of reaching 350 tons per day as soon as possible. "A telegraph line has been in operation for some time and a telephone line is just being completed."

Early Struggles

Osborn went iron-prospecting in Canada during August (as he did every year) and arrived home to learn of possibly the quarry's first industrial accident and to find Martin in need of cash. He told Fitch in a letter August 26: "He came in from the quarry and said he had shipped over $3,000 worth of stone this month and that he was dead broke." He said he had wired you but that you were in Chicago. He wanted $500 to use in paying wages and to settle with a man whose fingers were cut off at the quarry. I signed a note with him due September 15. When you pay him, will you please keep out that amount?" Fitch and Osborn's deal with Martin shielded them from liability for the injured worker, who later filed suit and won a judgment, but not large enough to hamper Martin's operation.[15]

As Fitch, Osborn, and Martin struggled to get the quarry producing and paying steadily—and Fitch ran his railroad, and Osborn pursued his iron prospecting—several Michigan newspapers reported that Osborn was fending off talk among friends and political allies in September that he should run for the U.S. Senate. Osborn may have declined to run for many reasons, but one might well be that the Iron Hunter was about to strike it rich. Osborn and several partners had an interest in an area he had studied known as Moose Mountain, near Sudbury, Ontario. In late October or early November, 1905, Osborn and his partners sold their interests to Canadian railroad builder Mackenzie and Mann for $1 million, of which Osborn received $250,000. *The Evening News* broke the story November 6, with the *Detroit Journal, Detroit News,* and *Detroit Times* all reporting it the next day.

Case Resurfaces, Briefly

Nearly a year after the quarry's false start, John J. Case sent Fitch a letter from Lima, Peru, dated November 3, 1905, containing a detailed bill for expenses "which I incurred on your account while examining and getting the quarry started." Case related that he was then engaged "to put in operation the large smelting plant for the Cerro De Pasco Mining Company."

The bill listed expenses for trips to Marquette, Houghton, Detroit, Duluth, and the Canadian Soo and supplies and time spent (78 days at $10 per day) totaling $1,096.75. The bill reached Fitch in early December and he relayed it to Osborn, who called it "a piece of presumption that hasn't even the merit of being exquisite."

"He agreed to do certain things in connection with opening the quarry on his own account," Osborn wrote. "It was really a gentleman's agreement and I imagine such a thing has strange significance to Mr. Case." Noting that Fitch "went ahead and built a spur track at a large cost," Osborn recalled their trip to the quarry during which they found two or three men "aimlessly awaiting orders."

"I think we will simply file his bill away," Fitch replied. "If he wants to bring suit against us we better let him do so."

The quarry was still struggling financially three months after Osborn helped Martin meet payroll in late August. In their exchange of letters about Case, Osborn and Fitch also discussed taxes and the company's health. "The Fiborn Limestone Co. is not particularly flush with money just at present and they owe me nearly $10,000," Fitch wrote, before asking Osborn to cover a tax bill.

Despite the tight cash flow at times, the quarry turned a profit for the year 1905. Its statement of operations December 31 reported revenue of more than $26,700 from the sale of nearly 55,000 tons of limestone and earnings of slightly less than $2,440, a profit of 9 percent.

Illness and Interest

As the quarry hummed along to open 1906 (a note in Osborn's

papers listed 108 cars of limestone shipped January 1–15), Fitch was left to oversee it for much of the first half of the year, after Osborn suffered an illness, described by one account as "overwork, his nerves and heart affected." His doctors ordered him not to see anyone or do anything that would make him tense. Eventually, treatment required a month-long trip to the Mediterranean and the Alps. Fitch relayed reassuring news in an April letter: "The quarry is doing well and we are getting inquiries from new quarters for the stone. It will work up into a handsome business."

Osborn and Fitch were still looking to sell their quarry, and in a May 22 letter Fitch reported he had given an option on the land "to one of the Merritt family of Duluth, Minn., for $75,000." The Merritts, who apparently wanted to buy the property and sell it to United States Steel Company, kept expressing interest but wouldn't commit to a deal, and Fitch refused to extend the option.

The quarry had become a village, and in a September 26 letter to Hadrich, Fitch noted he, Osborn, and Martin discussed founding a post office there. The application, filed in October, said the prospective post office would serve 75 people in the village and 125 people overall.

Fiborn Limestone Company's 1906 statement of operation showed revenue of more $60,800 on nearly 124,900 tons of limestone shipped and earnings of $9,060, a nearly 15 percent profit. The statement also showed $483 being paid "out of earnings in 1905" for a boarding house, indicating it had been built or construction was underway as 1906 ended.

A Post Office and Everything

The small hive of industry launched in 1904 entered 1907 with a boarding house and post office (officially established January 31 with Martin as postmaster), as well as a company store and five block homes for top employees and their families. A public school was built and operating by the fall, with up to a dozen students its first few years.

State geologist Alfred Lane returned to the area and reported: "The caves which I visited in 1901 ... were revisited in the spring of 1907, and have been opened out into extensive quarries. ... The shipment of the limestone ... runs up to 19 to 30 cars a day."

Martin, who could use timber from the property for quarry purposes at no charge, wanted to add a lumbering operation for profit, and offered Osborn and Fitch timber royalties of $15 an acre. The partners took their time considering the proposition, and eventually settled on $17.50. Martin built a sawmill and turned out rough lumber and shingles.

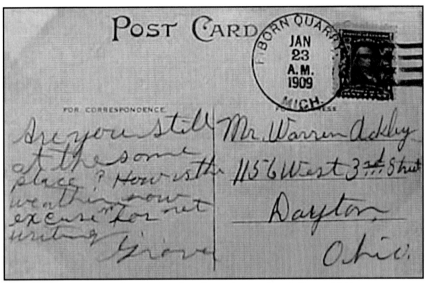

A post card mailed at the Fiborn Quarry Post Office in 1909.

Fitch declared three dividends in 1907. In February, he and Osborn divided $6,000 for the years 1905 and 1906. In September they divided $4,000 and in October divided $3,000. Fiborn Limestone would see annual profit jump nearly 24 percent to $11,194 from the previous year.

Despite the overall profit for 1907, business slowed greatly late in the year. A panic on Wall Street late that year led to a short but sharp recession. Fiborn's customers fell behind on payments, and some shut down completely. "The Zenith Furnace is shut down until March," Fitch wrote Osborn on December 4. "The Newberry Furnace closed yesterday so that all we have now is the Canadian furnace and the Copper Country furnaces. ... As long as the Canadian furnace runs we can manage to keep going but if it should shut down ... I do not just see how we could keep the quarry going." Algoma used about 300 tons of rock a day while other customers ordered only 200 to 300 tons per month.

Customers owed $8,000; Algoma nearly half of that. Fitch sent Martin to personally press for payment, and noted they owed Martin $2,000 but had only $1,700 in the bank. "The outlook does not seem especially good at present, but I suppose we shall have to accept the influence of conditions and abide them as everybody else does," Osborn replied. "The greatest concern I have is that the quarry shall not give you either worry or work. If it is going to give you any great trouble, I shall feel like advising you to shut it down at once."

As 1908 opened, Fitch and Osborn, whose busy life now included serving as a University of Michigan regent, saw yet another fleeting interest in the quarry pass quickly, this time from Rogers-Brown & Company of Chicago.

Fiborn Limestone's balance sheet worsened. In March, Fitch told Osborn the outstanding balance had grown to more than $9,940 and that Algoma's share had ballooned to $9,400 of that (the steel company had not paid for stone in December, January, or February). "Up to March 1st we owe Mr. Martin about $5,000 and we have a balance in the bank of $2,693." Fitch noted other customers' problems and concluded there would be no dividend until revenue improved. Despite that, Martin apparently was looking to increase the quarry's capacity. Fitch reported in April: "Mr. Martin went to some small place close to Chattanooga to examine a large crusher plant that has just been put into operation."

Business must have looked a little better to Fitch by May 1, when he issued Fiborn Limestone Company's fourth dividend, $1,000 each. And it must have looked a lot better by the end of the month, as he and Osborn had a deal to sell the property containing Fiborn Quarry for $100,000 to Algoma Steel's parent company, Lake Superior Corporation.

The Big One

The contract, dated May 20, granted "the option and privilege of purchasing the said quarry at any time within thirty days," payable with $25,000 down and three annual installments of $18,750. Samuel Martin, who leased the land from Fitch and Osborn's Fiborn Limestone Company, would continue to own the quarry equipment and lumber-mill operation.

Osborn recounted in his diary travelling to Marquette on May 19 to meet Fitch. "Got a quarry dividend of $3,000," he wrote. "Gave option to Lake Sup. Corp. at 100,000." After lunch at Fitch's home, Fitch and Osborn traveled on Fitch's private train car, stopping at the quarry on their way to Sault Ste. Marie.

It's not clear from Osborn's papers and newspaper accounts whether this deal stalled and was extended, or fell through and another one negotiated, but Lake Superior Corporation's purchase took a few more months to complete. Finally, in late October—just after Osborn launched another campaign for governor—they closed the deal. "Mr. Fitch in town yesterday," Osborn wrote in his diary October 20. "Have agreement on quarry deal." He later noted the deal closing on November 2.

On November 3, *The Evening News* reported: "A deal is pending whereby it is expected that the Algoma Steel Company will purchase the limestone quarry now owned by Messrs. Fitch and Osborn. ... The terms of the proposed transfer of the property are not made public."

Fitch and Osborn had done well with their property, making slightly more than $28,000 in profit from 1905 through 1908, and taking $19,000 of it in dividends by the end of 1908.

One Last Loose End

Eleven years after he and Osborn explored the caves and carried out samples of limestone, Charles B. Smith, now in Saskatchewan, apparently heard of the Fiborn deal. He wrote to Osborn in March 1909, saying Osborn had spoken of an interest in the limestone of a thousand dollars or more, and implying he had been cheated. Osborn was in Europe at the time, arrived home in July and complained "I do not like the tone of your letter" in his reply. Osborn recounted offering to pay Smith at the time of the trip for his troubles and Smith refusing. "I then said that I would have to pay you later. This statement did not carry the understanding that you were to have an interest in the lands, or in the stone, or in anything else."

Smith's rambling four-page handwritten rebuttal of August 2 reciprocated Osborn's umbrage over tone and denied that the latter offered to pay him that day. Smith insisted "at the time you said my interest was worth one thousand dollars but would be worth more later," and that others heard Osborn say it. "Showing you and the public that I am no grafter is [*sic*] that you place a draft in the Bank of Montreal Regina to my credit for $1,000 with interest at 6 percent compounded from date of your partnership with Wm. Fitch to present date. Then Mr. Osborn you and I will be through."

Osborn spent August prospecting in Canada as usual while Smith's letter languished in his office, and the weeks in the backcountry appear to have mellowed him. "I am sorry that a misunderstanding has arisen between us," he wrote September 2. Still quibbling over details and insisting he did offer Smith $100 for his troubles that December day, Osborn proposed sending him $500 in a bank draft. Smith did not immediately accept, and Frank Perry (owner of the lumber camp where Smith worked when he explored the cave with Osborn) stepped in as mediator. He basically brokered the same offer Osborn proposed, and on October 27, Smith wrote, "I give you full power to settle this question any way you wish and will be satisfied if I don't receive a nickel." Perry wrote to Osborn on November 5 seeking a bank draft for $500, Osborn sent it and apparently never heard from Smith again.

After selling the quarry property, Osborn continued his run for governor and won, serving through 1912. As governor, he championed a reformist agenda culminating in the adoption of a workman's compensation law. He also endorsed and persuaded the state Legislature to place a proposal granting women the right to vote on the 1912 ballot, which lost by only 762 votes. He left office and went on a world tour, saying he wasn't interested in running for governor again (he had pledged to serve only one term), but ran again unsuccessfully for the Republican nomination in 1914.[16]

Fitch retired as president of the DSS&A in November, 1911, remaining with the company's board. He also was elected vice president and a director of Wisconsin Central in 1912.[17] He died at age 78 on September 16, 1915, at his home in Marquette after several months' illness. In Fitch's honor, DSS&A officials ordered railroad employees to stop all work for two minutes at 2:00 P.M. on Saturday, September 25, according to the *St. Ignace Enterprise*, which called him "one of the best known railroad men in the middle west."

Fitch and Osborn remained close after they sold Fiborn Quarry. As Fitch grew increasingly ill in the summer of 1915, Osborn wrote twice during his usual August Canadian prospecting travels, asking about his friend's health. When Fitch died the night of September 16, Osborn again was in the north woods and unreachable. He missed the funeral, but later wrote to Fitch's daughter, "I loved him as a son loves a father."

Osborn continued to dabble in politics, write, travel, and manage various business interests, dividing his time between his later homes on Duck Island in the St. Marys River and in Georgia. He died at age 89 in 1949 in Poulan, Georgia.

Meet the New Boss

Algoma Steel was chartered in April, 1901, by entrepreneur and promoter Francis Clergue, who came to Sault Ste. Marie, Ontario, in 1894 amid long-running attempts to develop hydroelectric power at the falls in the St. Marys River. Clergue and others envisioned an industrial hub fueled by abundant, low-cost electricity. After buying the Canadian city's stake in the then-stalled power project, Clergue won a 20-year franchise on water power, incorporated electric utilities, a pulp-and-paper company, and other businesses including Algoma Steel. The holding company that owned it all, Consolidated Lake Superior Company, was launched with $117 million in capital, raised largely through Clergue's salesmanship.

The Canadian and Ontario governments sought to foster a steel industry using iron ore from deposits in the region. The promise of cheap hydroelectric power and low-cost shipping by boat made the Soo an attractive choice to some, despite the fact the city wasn't near large supplies of iron ore, coal, and limestone needed to make steel.

Clergue, in a move typical of the company's early missteps, built blast furnaces before smelters, forcing the plant to use pig iron purchased from other primary producers, which was much more expensive than iron ore from mines. The plant produced steel rails for Canadian railroads, a business which grew substantially during the 1890s. Though Clergue boasted in April, 1902, that the Algoma works employed 7,000, production problems and flawed batches of rails led to contracts being canceled. Clergue shut down the plant in December. After running in fits and starts, the plant closed again in September, 1903, when its parent company couldn't cover the payroll and filed for bankruptcy protection. By February, 1904, Clergue reconstructed the holding company as the much leaner Lake Superior Corporation, headquartered in New York with $40 million in stock and $13 million in bonds. Clergue, however, began to lose his hold on the company and was replaced as president in April, 1904, by C.D. Warren. Clergue remained as an adviser and a member of the board of directors, but lost that post in 1908.[18]

From 1904 to 1913, a growing Canadian economy, new transcontinental railways, and government tariffs protecting domestic

33

steel production led to a few years of steady output and profits at Algoma. An economic panic in 1907 caused production cuts in 1908 and 1909, but the company remained profitable and demand picked up steadily after that until 1913.[19]

Along with Fiborn Limestone Company, Lake Superior Corporation bought Cannelton Coal Company of West Virginia in 1910 and made both subsidiaries of Algoma Steel. A board of directors oversaw Fiborn Limestone; company officers included a secretary, a comptroller and a treasurer. (Balance statements from the 1920s listed a 12-member board, whose members were from the Canadian Soo, Toronto, New York, Philadelphia, and other eastern American cities.[20])

Samuel Martin continued to own the equipment and run the quarry. By or during 1910, he incorporated S.B. Martin Company, capitalized at $225,000.[21] Fiborn Limestone Company planned to spend $150,000 in 1910 to expand the quarry's production capacity from 1,000 tons a day to 1,000 tons per hour, and its storage capacity from 900 tons to 4,000 tons, The *Michigan Manufacturer* magazine reported on March 19, adding "the limestone supply is practically unlimited, covering an area considerably over 680 acres." *Cement and Engineering News* that same month reported that S.B. Martin Company bought a "Mammoth McCully crusher," along with a conveyor, screens for sorting crushed rock, elevators, and other equipment.

S.B. Martin Company signed a five-year lease for the property with Fiborn Limestone, the St. Ignace *Republican-News* reported August 13, adding that Martin planned to "commence operations on an extensive scale." He and the company bought three new locomotives and other equipment during 1911, according to retroactive reporting by *The Evening News* in 1913.

A Company Town

Fiborn Quarry employed 46 men, not counting boarding house staff, and 74 people lived in the village when the 1910 census was taken in April. Some 50 people were listed as boarders in household No. 74, headed by quarry manager Samuel B. Martin, which consisted of the boarding house and at least two log bunkhouses. Some boarders were listed as woodsmen or sawmill workers, likely employees of Martin, who by then owned 240 acres immediately adjoining the quarry property and ran his sawmill there. Of the five family residences in the census, three were occupied by couples with no children.

The workforce was a mix of Michigan natives of several European strains, a scattering of men from Ohio, Indiana, Pennsylvania, and New York, and nearly two dozen immigrants from Canada, Finland, Bulgaria, Sweden, the British Isles, and France. The boarding house employed a housekeeper, cook, waitress, and two dishwashers—all women. One was from Indiana, the others from Finland, Sweden, and England. Two had children, and apparently were separated by employment from husbands and other children.

Tidbits of life at the quarry were infrequently published in the St. Ignace *Republican-News* in the early years of the decade, and in 1910 they mainly evoked the adage "no news is good news." Or maybe it was just a particularly bad year. Along with stories of fires and explosions, the drowning death of a seven-year-old boy who "fell from the pump house into the lake," and the death of a four-month-old boy from "cholera infantum" were reported during the summer. Otherwise, business visits by quarry employees or notes about residents moving to and from the quarry made up most of the correspondents' dispatches from Fiborn Quarry and nearby villages.

Dangerous Business

Martin's sawmill business made news when it burned to the ground under suspicious circumstances. The *Republican-News* of July 23 relayed word that a "discharged employee had boldly told Mr. Martin that he would burn him out, and it is thought that the fellow carried his threat into execution." Noting there was no regular night watchman on duty, the report said a woman who lived nearby was up with a newborn around 12:30 A.M. and saw no fire, but the "whole mill was wrapped in flames" at 1:15 A.M. "Mr. Martin's hands and arms were badly burned in efforts to stay the flames." Along with the mill, stockpiles of shingles, lumber, and uncut timber were destroyed.

Tragedy struck the quarry a few months later. On December 13, workers Eric Erickson and Naffre Nelson died when dynamite exploded prematurely while they were packing sand in a blast hole. According to the *Republican-News*, the Mackinac County coroner and sheriff went to Fiborn Quarry, empanelled a jury, and held an inquest. The jury returned a verdict supporting management that the blast was accidental and had no obvious cause.

Only two months later, another accident killed one worker and injured seven, five of whom were taken by train to a hospital in Sault Ste. Marie. The boiler powering the crusher "let go" about 4:30 P.M. on Friday, February 17, 1911, according to an account published the next day in *The Evening News*. Scalding accounted for many of the injuries. John Mudila, the man killed, had come to the U.S. from Finland about a year earlier. He reportedly had a wife and children still in Finland, and was planning to send for them soon. "The boiler and boiler house were wrecked, and the loss will be in the neighborhood of $10,000," the paper reported. "The work of installing a new boiler and otherwise repairing the damage began this morning and it is expected that everything will be running as usual by Monday."

As with the 1910 dynamite explosion, whose victims also were of Finnish descent, the Mackinac County coroner convened an inquest, this one in St. Ignace immediately after the accident. "No cause could be assigned for its blowing up," *The Evening News* account said. A fireman, who escaped injury, testified he had checked the boiler's water level shortly before the blast and it was two-thirds full. The boiler "was considered in first-class condition."

Martin's New Venture

As 1913 opened, Martin sold his quarry and lumber operations to Algoma Steel. "A deal has been consummated whereby the Algoma Steel Company, Canadian Soo, becomes the owners of the limestone quarry, plant, and equipment formerly owned and operated by S.B. Martin, at Fiborn," *The Evening News* reported January 12. "The purchase also includes 240 acres of land immediately adjoining the property." Martin formed Martin International Trap Rock to run a quarry in Bruce Mines, Ontario. "The company will have offices here and at Bruce Mines with

S.B. Martin as manager," the paper reported. (Martin would return to the area in 1916 and start a new quarry in Ozark, just south of Trout Lake. As he did at Fiborn Quarry, Martin became postmaster of Ozark and moved the post office to his quarry. Eventually, he sold Ozark Quarry to Fiborn Limestone's parent company, Lake Superior Corporation, and headed to New Jersey, where he continued in the quarry business until his death in 1934.) Martin was succeeded at Fiborn Quarry by 30-year-old fellow Ohio native Oliver Peter Welch, who had worked at quarries in his native state as well as mines on the West Coast and in Alaska before finding his way to northern Michigan.

Steelmaking in Canada generally and at Algoma in particular hit record levels in 1913[22] but by then the U.S. and Canadian economies were falling into recession, bottoming out in late 1914.[23] In December, 1914, Algoma Steel reported its plant was operating at only 40 percent of capacity.[24] Fiborn Quarry likely was running at similar levels.

While workers quarried limestone at Fiborn Quarry and milled rails at Algoma, and while their employers grappled with the recession, the assassination of an obscure member of European royalty on June 28, 1914, was to touch off war across much of the continent, and its economic impact would reach even isolated communities such as those in the Upper Peninsula.

World War I Boom

Since the United States didn't enter World War I until April, 1917, the Great War's only early effect on Algoma Steel Corporation and Fiborn Limestone Company was presenting Algoma the opportunity to diversify production and rely less on railroad business, especially as demand for rails began tumbling after Canadian steel production hit a record in 1913.[25]

By one account, Fiborn Quarry in 1914 operated ten hours a day, six days a week, shipping about 30 cars, or 1,500 tons, of limestone a day, at a profit of 50 cents a ton or $750 a day. [26]

The hustle and bustle of trains carrying limestone and lumber up and down spurs to junctions on the Duluth, South Shore & Atlantic line could and often did cause wild fires. A wave of fires in 1914 apparently led to a crackdown by a local game, fish, and fire warden that got the St. Ignace *Republican-News* into a spat with him and again landed Fiborn Quarry on the front page.

Declaring in the top headline "He was too humane," a front-page article in the June 6 issue detailed how quarry master mechanic Charles Collins "was hauled into Justice Reagan's court on Monday and fined $5 for having no spark arrester on the engine." The fourth deck of the headline declared Fiborn Quarry's master mechanic "Fell Victim to Lunk-Headed Official Zeal and Fined as His Reward."

Collins claimed in court that after being (rightfully) confronted by Deputy Game Warden Frank W. Nelson over the lack of a spark arrester (a web of wire in the smokestack that prevents burning embers from escaping with exhaust), he promised to rig one before running the locomotive again, then returned to the quarry where, a day or two later—and before he could obtain the proper wire—a woman broke her ankle and needed to go to a hospital, so he risked a trip to the South Shore line. Collins said he arrived at Fiborn Junction and Nelson just happened to be there, and cited him.

"Comment upon the case is superfluous," the newspaper commented. "When will piffling be finally cleaned out of that colossal graft, the state game, fish, and fire department of Michigan?"

However, tucked atop the second column of page two in the

following week's issue, under a tiny headline whispering "Contradicts Collins Flat," the *Republican-News* ran a five-paragraph statement from Nelson, with an introduction likely meant to prevent a libel suit, that said "George Collins' pathetic story to Justice Reagan, as narrated in last week's *Republican-News* ... was a deceptive misapplication of a previous incident."

Declaring "I can prove by train men and others that this woman was brought out May 18"—almost two weeks before the confrontation reported in the June 6 story—Nelson went on to describe inspecting all three of Fiborn Quarry's locomotives and finding "none of them had any spark arresters or ash pans" as required by law.

"I ordered the locomotives sidetracked for repairs May 22, and warned Mr. Collins not to use any of them until repaired. But, contrary to my orders, locomotive No. 106 ran out to meet the South Shore train on the evening of May 22, also on May 27 and 28. This is what Mr. Collins paid a fine for and not bringing out the woman he speaks of, on May 18."

Fiborn Quarry wasn't the only violator. A June 11 *Enterprise* article on the issue of fires caused by trains, noted: "Mr. Nelson says that of the 19 engines inspected this season, only one had complied with the requirements of the law. He says further that over half of the forest fires in the past have been caused by sparks from engines."

Romance and Roadsters

Possibly because the quarry was busy and prospering in 1914, the occasional dispatches from its anonymous correspondent to the *Republican-News* seemed much more upbeat than those in 1910, if not as informative. A dispatch published September 26 opened cryptically with the question: "Who said wedding Bells will soon be ringing?" Talk of marriage continued with something of a classified ad. "Wanted: a wife, by a middle-aged man; owns seventeen lots in Detroit across from P.O.; four diamond rings, horse and cac. [*sic*] Good looking, intelligent ladies only need apply. — P.B. Fiborn Quarry."

The September 26 column also makes the first recorded mention of automobiles at Fiborn Quarry, and even that item carries on the week's romantic theme: "Our young folks are thinking of going for another gasoline car ride next Saturday night. How about it girls?" Soon after, the *Enterprise* Fiborn Quarry column reported: "A number of young folks enjoyed a pleasant ride to the junction Wednesday evening. They were entertained beautifully while there, dancing being the chief amusement."

The growing use and importance of cars was evident in a February, 1916, *Enterprise* item about a local resident getting a county road job, which noted "the boys from Fiborn Quarry are anxious to get through with their autos this spring." Four months later, the paper reported that Mackinac County authorized Hendricks Township "to raise $5,000 through loan or bonds to complete the road to Fiborn Quarry."

As much as residents may have felt life changing and becoming more modern in 1916, some problems they faced weren't much different than those faced by Native Americans generations earlier. For example, making sure your food doesn't become the local wildlife's food, as in the case of "A bear burglar" reported by the *Enterprise* on July 20: "It is no uncommon sight for people residing at Fiborn Quarry to see a bear strolling in the woods or to hear them at night about their doors in search of something to eat. Last week a carload of supplies arrived at the quarry and before its unloading was completed it was dark and the work was left to finish up the next morning, which proved to be a case of bad judgment. The car door was left slightly ajar and sometime during the dark hours a bear noticed this and proceeded to make an opening sufficiently large to let him in. Among the remaining contents of the car was a case of raisins and a case of butter. The raisins looked mighty good to his bearship, who 'licked the platter clean.' Finding no bread for the butter he evidently decided to wait until he made another survey of the neighborhood in order to secure a supply, for he took the case of butter a considerable distance away from the car and buried it in a sand bank. The night was warm and the dripping of the butter along the way to the cache led to its discovery. An armed guard has kept up a watchful waiting throughout the silly hours of the night ever since, but so far the bear has not made any attempt to recover his buried treasure."

The World Intrudes

In 1917, the war in Europe became more than a story "over there" and a vague upward tug on industrial and agricultural output. Congress voted April 6 to declare war on Germany. Draft registration began in June, followed by a second round in August.[27] A third round later sought to find people with special skills such as construction, for non-warfare work like rebuilding and logistics.

The *Republican-News* reported on June 17 that 780 men in Mackinac County had been registered, with more than 300 qualifying for exemptions. In July, when the first drawing of draft numbers took place, Mackinac County's quota was estimated at 58 men, and included at least five from Fiborn Quarry.[28] Nine more Fiborn Quarry residents were called in August.[29] The first contingent of Mackinac County draftees, six total, left St. Ignace on August 31 for Camp Custer near Battle Creek

41

with a sendoff that included flag-waving citizens, speeches, and a band. None of the six were from Fiborn Quarry, as were none of the 26 men in the second contingent which left September 21, and none of the 19 in the third contingent on November 23.

At least two Fiborn Quarry men eventually served in the military, Henry Pope and Angelo Perone, according to lists of 312 men who had been drafted or enlisted which appeared in both St. Ignace papers in mid-December 1918. Pope doesn't appear in any records or coverage of Fiborn Quarry other than a couple of lists. Perone, who came to the United States from Italy in 1911, appears in the 1920 and 1930 census rolls, and was mentioned occasionally in St. Ignace newspapers.[30]

Fiborn Quarry Pitches In

After the U.S. entered the war, bond drives to finance the war effort and campaigns to raise money for causes such as soldiers' aid tugged on Americans' finances. Fiborn Quarry Superintendent Oliver Welch, like most of his peers in American business, was a supporter of and participant in these drives, and likely made sure his employees were too. As the *Enterprise* reported a day after July 4 celebrations in 1917: "Fiborn Quarry is on the map in the war game and every man employed there has gone on record as desiring to do his bit for the Mercy fund and the war. General Superintendent Welch on Saturday sent to the secretary of the Mackinac County chapter $120 contributed to the $100,000,000 fund and the county chapter of the Red Cross by the employees of the Fiborn Limestone Company Mr. Welch himself contributed $25." A list of contributors contained 23 employees.[31]

Four Liberty Loan bond drives were held during World War I to help finance America's mobilization. As with the Red Cross drive, Superintendent Welch pressed Fiborn Quarry into action. As the *Enterprise* reported April 25, 1918: "O.P. Welch, superintendent of operations at Fiborn Quarry and in charge of the Liberty Loan drive for that place, was the first to send in his report ... stating that every man but one at the quarry had subscribed for bonds and that the one holding out would be on the list of buyers before the close of the following day. $2,200 was the amount subscribed. The *Enterprise* believes that this record will stand throughout the district and Mr. Welch is to be congratulated for his patriotism and his generalship as head of the drive in his section."

The quarry village in 1917 and 1918 was probably very little changed since Samuel Martin built the boarding house, block houses, and company store. Under Algoma Steel Corporation ownership, Fiborn Limestone Company began adding cottages for workers with families, but through the war years, Fiborn Quarry remained a community of mostly of single workers. The *Republican-News'* Fiborn Quarry column chronicled the comings and goings of village residents, new and departing employees, and reflected their everyday concerns, such as these highlights from August 11, 1917: "Everybody is busy picking red raspberries. ... The frost did great damage to everybody's gardens. ... Everybody expects to go to Caffey to the Hard Luck dance tonight."

Post-War Bust

The war caused demand for steel to be made into munitions rather than rail. By 1917, Algoma Steel had dismantled some of its rail mills and was producing more steel ingots. Despite the fact the steelmaker still relied heavily on outside purchases of coal, production and profits grew until peaking in 1918 at slightly less that 500,000 tons of steel and nearly $7 million for parent company Lake Superior Corporation. The war's end in 1918 pulled the rug out from under economic growth across North America. By mid-1919, Algoma's rail mill was operating at only 46 percent capacity and the company reported a marked decline in profits.[32]

Fiborn Limestone Company showed net earnings of $35,317 for the first half of 1918, but that fell to $18,996 in the first half of 1919 and $6,386 in the first half of 1920.[33] Capital improvements likely ate into profits. The company bought a $1,600, 18-foot lathe and shaper for the machine shop in 1919, a $2,900 steam drill in 1920, and an $8,500 43-ton locomotive the same year. Electricity would come to the quarry in 1921 with the purchase of a $1,050 10-kilowatt electrical generator.

Change and unrest followed the war's end. Women in Michigan gained the right to vote a year ahead of nationwide adoption of the 19th amendment to the U.S. Constitution. Ratification of Prohibition, the anti-foreigner Red Scare, and widespread strikes and labor unrest played out through the recession year of 1919. The wave of industrial unrest seems to have reached even tiny, isolated Fiborn Quarry. The *St. Ignace Enterprise* on November 13 reported "quarry Superintendent Welch, on a business visit to the town, said that there is not now nor has there ever been any labor trouble at the quarry. ... It was found necessary during the summer to discharge a number of labor agitators who endeavored to make trouble, but were unsuccessful. These men have applied for reinstatement with promises to be good in the future if given their old jobs back."

Welch left Fiborn in early 1920, before the census was taken in April. During his time as quarry manager, he had developed business interests in St. Ignace, and by the time he left Fiborn had become a partner in the city's upscale Dunham House Hotel.[34] Valentine Hemm took over as superintendent at the ripe old age of 24. Like Welch, Hemm was from Ohio, to which his parents had immigrated from Germany. He was single at the time and lived at the boarding house, as did another future superintendent, Harry Myers.

The 1920 census showed a smaller quarry operation, with only 16 workers (three of whom were managers) and a total of 32 village residents. Thirteen workers lived at the boarding house, while 19 people lived in four units of family housing. As in 1910, Fiborn's workers included immigrants from Italy, Finland, Russia, and elsewhere.

The postwar recession all but shut down Algoma, and Fiborn Quarry with it, for much if not most of 1921 and 1922.[35] Fiborn Limestone Company balance sheets reported a loss from operations of $5,471 for the first half of 1921, followed by losses of $7,986 and $9,440 for the same periods of 1922 and 1923. "We wish Mr. Hemm would wake up and start the quarry so Fiborn won't die," lamented the *Republican-News'* Fiborn column of April 23. (Despite that, the column noted: "The folks from here had a good time at the dance last Saturday.")

Signs of Life

Finally, early in 1923, the business stirred. "H.V. Hemm, superintendent of the Ozark and Fiborn quarries, which have not been operating for some time, left Thursday for the purpose of placing the quarries there in commission again," reported the *Enterprise* on February 23 (misprinting V.A. Hemm's initials). "He says a small force will be employed for the present but expects soon to be operating at both Ozark and Fiborn at capacity."

Well into spring, the small hive of industry founded by Fitch and Osborn and built by Martin was reborn, reviving regular correspondence published in St. Ignace newspapers. The *Enterprise's* Fiborn Quarry column reported in April: "The quarry is as busy as Ford's factory." Maybe most importantly for the winter-weary residents of Fiborn, the snow plow arrived, according to the *Enterprise* correspondent. (Plows were horse-drawn until the late 1920s, based on former residents' recollections.) As the quarry and village emerged from an Upper Peninsula winter's snowfall by mid to late April, spring fever and its evil twin, cabin fever, surely ran high. Major League baseball's 1923 season had just opened, complete with The Babe swatting home runs from day one in the spanking new House That Ruth Built.[36] Why let some leftover snow and not-quite-warm temperatures prevent red-blooded Americans from pursuing the national pastime? "Everybody played ball Sunday," the last item of the April 26 column began. "Mr. Hemm umpired the game, while Mr. Fyfe (the bookkeeper) acted as score keeper. No spectators were allowed as the game was played down at the quarry."

As strange as it may seem in an isolated area with nothing but a few canoeable streams and ponds for navigable waterways, Hemm managed to build a speed boat at the quarry in 1924. The *Enterprise* on July 31 announced: "V.A. Hemm's new speed boat, built at Fiborn, arrived here by rail Tuesday and was launched in the bay in front of the Hotel Northern yesterday." No details about the boat were reported. The speed boat may have been a parting gesture on Hemm's behalf. By then, he had become partners with his predecessor, O.P. Welch, in the Hotel Northern, and left Fiborn late in 1924 or early in 1925.[37]

Harry Myers, 32, succeeded Hemm as superintendent. An Indiana native, Myers began working as an engineer at Fiborn in 1920. Myers' wife, Marcella, and four children moved from Rexton into one of Fiborn's block houses after he became superintendent. (Marcella died after an illness in 1929 and the children moved back to Rexton. Myers resumed living at the boarding house.[38]) Myers remained superintendent until about the time the quarry closed. Another engineer and future superintendent arrived in 1924, locomotive engineer Lynn Brockway. His previous railroad experience made him "dean of the engineers," said a former village resident. Lynn worked for the Michigan Central Railroad before Fiborn, and for a Pennsylvania railroad before that. Brockway brought his wife, Lillian, and sons, Harry and Bernard, to Fiborn in 1926.[39]

As the 1924–1925 school year was about to get underway, the teaching job at Fiborn was vacant and the district couldn't fill it, so the newest company wife was pressed into service. Vera McEachern had recently moved to Fiborn from Sault Ste. Marie, Ontario, with her husband, Donald, the new accountant in charge of properties for Fiborn Limestone Company, and their two children.[40] As well as being able to teach, Vera was an accomplished pianist and organist. Don was 25 and Vera 21 when they married in 1913 in the Canadian Soo. Their daughter, Jean, and son, Keith, were born there.

Keith, interviewed many years later, recalled moving to Fiborn during the summer of 1924. "I was about five years old. When we moved in there, I remember they took us in with a horse and buggy. ... We had [a] big grandfather clock sitting on the seat beside me."

In addition to Don's accounting and administrative duties, he and Vera were to run the company store, which also housed Fiborn Quarry's post office. Vera ran the store and became, as the 1930 census roll called her, postmistress. That a 31-year-old mother of two would take on the load of teaching elementary school when already assigned a store and post office may seem like an enormous burden, even by modern standards, but Vera McEachern was just warming up.

Recovery, Then Recession

After the dismal years of the first half of the decade, Algoma Steel steadily increased production during the late 1920s.[41] After the string of losses through 1925, Fiborn Limestone Company earned $7,480 from operations in the first half of 1926, $10,882 in the first half of 1927, and $17,246 in the first half of 1928.[42]

While Fiborn Quarry's workforce grew modestly from 16 employees in 1920 to 22 in 1930, according to the census, the village population ballooned from 32 residents to 87, and the housing balance shifted strongly to families, particularly large ones. In 1920, only two of the four-family households listed school-age children, with a total of 19 people living in family housing and 13 people living in the boarding house. In 1930, only six workers lived in the boarding house, while the other 81 village residents were divided among 16 families (and included six non-family lodgers). Automobiles, improving roads, and reliable county snowplowing made single workers more mobile, and led to a shrinking population of single men in the boarding house. It may be that the company hired more family men for a remote setting, counting on lower turnover.

As 1927 opened, the quarry was shut down for repairs, a typical practice for a month or two in the winter. The St. Ignace *Republican-News* Fiborn column of January 8 speculated: "Suppose the men will be going to camp to cut cedar now."

As production and profits increased through 1928, Fiborn Limestone Company bought a 30-ton rebuilt Vulcan locomotive for $3,000 early in the year to replace a damaged 19-ton Davenport, a $7,000 Erie B2 steam shovel (for stripping soil and vegetation from limestone) in August to replace a Bay City crane model that "failed." In May, 1929, they purchased materials to build a block building adjacent to the machine shop to serve as headquarters for the superintendent and accountant, and storage for spare parts. The authorization for $2,426 to build the office noted, "We propose to utilize the present office building, formerly a hen house, by replacing it on the bank for a workman's cottage." (Interviews with quarry residents many years later indicated at least one employee did indeed live in a small shack resting on a ledge on the bank near the boarding house.) The 56- by 22-foot office included 12 windows so its occupants could survey almost the entire quarry works. Fiborn Limestone Company also approved $1,230 to build a Class A dynamite magazine—apparently the quarry's first regulation storage building in nearly 15 years of operation. "Dynamite is at present stored in small wooden building in pit alongside railroad storage track," the authorization form noted. "The erection of a proper magazine, to be located outside the pit, is recommended solely on the grounds of safety both in respect to workmen and plant."

The company also spent nearly $4,000 on four cottages from the nearby Wilwin Lumber Company in late 1929, a move that indicated confidence in the quarry's future. But the economic boom of the mid-late 1920s, much of it driven by rising stock prices, was beginning to unravel even as the Wilwin cottages arrived in sections on horse-drawn wagons. On October 24, several days of frantic selling and plunging prices began, peaking on Black Monday, October 29. Runs on banks and bank failures followed, slashing investment and industrial production, bankrupting companies, and wiping out the savings as well as paychecks of millions of people.

The quarry and village clung to life as the Great Depression threw Americans and Canadians out of work in huge waves. Unemployment among nonfarm workers in Michigan rose to 20 percent in 1930, then 29 percent in 1931, 43 percent in 1932 and nearly 50 percent in 1933.[43] "Labor conditions at the quarry are still very dull, with little prospect of change," reported the *Newberry News'* Epoufette correspondent on September 5, 1930.

Vera McEachern, after serving as a teacher at Fiborn and school board member while regularly playing piano and organ for church services and special events, was elected to the Mackinac County Red Cross' board of managers in February, 1932, and became increasingly active in efforts to help the hundreds of county families thrown into poverty by the Depression. In May, the *Republican-News* reported that the Red Cross was sending a truck loaded with 3½ tons of flour to four isolated communities including Fiborn, with Vera in charge of distributing it there. In 1934, she was elected chairman of the Mackinac County Red Cross, and also became active in the Upper Peninsula Association of Welfare Workers.[44]

While many village residents hung on to their rent-free homes and hoped for better days, others left. Workers without families had little or nothing to stay for; the boarding house closed about the same time as the school, in 1935.

Parent Company in Crisis

Ultimately, Fiborn Quarry's fortunes were tied entirely to the fortunes of Algoma Steel and its parent company, Lake Superior Corporation. Modern experts on manufacturing and supply chains might hold up Fiborn Limestone Company and Algoma Steel as a case study in the pitfalls of exclusive-supplier arrangements.

The industrial collapse after 1929 led to "a severe depression in all lines of business," Algoma said in its 1931 annual report. The steel company reported a net loss of more than $800,000 in June, 1932, and Lake Superior Corporation placed Algoma in receivership. During 1932–1933, the company operated one blast furnace and one battery of coke ovens, and by October of 1932 the plant's workforce was down to 332. In May, 1933, Algoma cut wages from 30 cents an hour to 27 cents an hour.[45]

Production did pick up in 1933 and 1934, and the steelmaker reported a profit of more than $860,000 that year. The plant went from operating only 84 days in 1933 to 325 days in 1935, but still was with only one furnace in use.

In 1935, Lake Superior Corporation and its Algoma Steel subsidiary underwent a wrenching overhaul under new majority owner James H. Dunn. He had been acquiring stock and bonds since the early 1900s, had made an unsuccessful bid to control the company in 1908, and proposed a restructuring in 1927, but with no success. Dunn held 72 percent of the company's stock and most of its bonds by the mid-1930s, and again proposed a restructuring plan, which his fellow bondholders approved in February, 1935. He became president in May, arranged a foreclosure sale of the old company, set up a new Algoma Steel Corporation with himself as president and chairman, and reorganized the company's financial structure, which included suspending the dividend on shares of common stock.[46] Under Dunn's leadership, Algoma would look to cut costs, which didn't bode well for Fiborn Quarry's future.

Working at the Quarry

Machines, technology, and people came and went, but the essential tasks remained the same throughout Fiborn Quarry's history: drill, blast, load, haul, crush, sort, and haul again.

Production typically slowed in winter to the point the quarry would shut down during the snowiest months and workers would repair and upgrade equipment. "They'd build a stockpile over at the steel plant so that they could be down in case the weather got so bad [the quarry] couldn't operate," said Keith McEachern. "We didn't have any snow plows. So when the snow got over about a foot, too deep, you couldn't operate." Snow melt in spring sometimes flooded portions of the quarry and held up production. "[Water] would be all down around the bottom of the bins, at the plant, the screening plant, as well as out in the quarry itself," he said.

Besides snow and ice and the fortunes of Algoma Steel, forest fires occasionally interrupted production. "Whenever there was a forest fire and [the state Conservation Department] needed help, they'd call on these guys to go and help to put out the fire," McEachern said.

The superintendent ran quarry operations while the bookkeeper took care of payroll and accounting, reporting to an executive at

The quarry works, including the railroad shed, power house, crusher and sorting bins. From a collection of photos taken in 1924 by quarry worker Aaron Thompson and preserved by Bernice Hood of Hendricks Township.

Algoma Steel rather than the superintendent. When limestone was being crushed and shipped, the bookkeeper worked in the 12-window office along the railroad tracks, recording every load shipped. Machinists, engineers, steam-shovel drivers, and a host of laborers supplied the muscle to make and move crushed limestone. A laborer might find himself crawling through the dark bowels of the boiler works to clean out dust and debris, perched in a bird's nest of a seat on the bouncing boom of a steam shovel, laying and then pulling up steel rails, shoveling coal into a roaring fire, hopping off a "dinky" and running ahead to throw a switch, fueling and watering locomotives overnight, or trying to free a jammed boulder in the crusher with a long steel rod or, occasionally, a stick of dynamite.

Algoma paid 30 cents per hour to its laborers in Sault Ste. Marie.[47] Fiborn Quarry wages likely were similar: Don Stokes, who worked at the quarry starting in 1929, said he and a brother made 32 cents per hour. Six 10-hour days was a typical workweek in American manufacturing through the early 1900s. At 30 to 32 cents an hour, Fiborn Quarry laborers made $18 to $19 a week.

The closest thing to a fringe benefit may have been free steam-cleaning of work clothes, thanks to the quarry's boiler plant. "They had a barrel there that we could just drop a hose into it and steam our clothes clean," recalled Clyde Ball, who worked at Fiborn Quarry in the 1930s. "And boy, it would take every bit of grease and dirt out of those clothes."

As hard as the jobs at Fiborn Quarry were, they were valued by those who held them in the hard-scrabble world of the U.P. in the 1920s and 1930s. People often were willing to work hard just to get one. Merton Carpenter, who lived at Fiborn Quarry as a teenager (his father rented a Fiborn Quarry cottage while working at the nearby Wilwin lumber camp), sought a job at the quarry in 1934, while living in Trout Lake. Carpenter said he walked back and forth between Trout Lake and Fiborn Quarry six days in a row, urged on by his friend Ball, who assured him he would get hired. Finally, Carpenter said, superintendent Harry Myers singled him out from a group of men hoping to get hired for a day's labor. "I'm tired of seeing you here every confounded morning," Carpenter quoted Myers as saying. "I'll put you to work so I don't have to look at you every day"

Drill and Blast

Horsepower was important in the quarry's early years. A blacksmith setup, including a small forge, was on John Conlin's original shopping

Steam-powered drilling rigs used to drill holes in which workers placed dynamite.
(Aaron Thompson collection)

list. Aside from the steam drill Fitch borrowed in 1905 to get quarry operations started, drilling during the quarry's early years involved a horse-drawn platform on which two workers would drive a blade-tipped metal shaft into the rock with sledge hammers while a third turned the shaft with a pipe wrench.

By 1920, tall steam-driven drills replaced the horse-drawn hand-powered type.[48] Crews drilled lines of 15 to 20 holes, typically 30 feet deep and 20 feet apart, a process one former quarry resident recalled taking up to a week. Water was pumped in to lubricate and loosen the rock. "They drilled and this 'lava' (mud) would come out from the rocks and just settle like a great big pancake, all the way around the drill," another former resident recalled. "Powder monkeys" placed explosives, then filled and packed the holes with sand. In the quarry's early years, long fuses set off the charges; later blasting caps and an electronic plunger did the job.

Blasting took place during afternoons so that if a charge didn't go off, workers could wait overnight to insure the dynamite wouldn't explode after a few hours, while they were nearby. Before blasting, a steam whistle would warn workers and village residents to take cover. Flying rocks often reached houses near the quarry, so families "used to go back in the woods," recalled one former resident. A teacher remembered pieces of plaster falling from the schoolhouse ceiling. Junior Shoemaker, who was born at Fiborn Quarry in 1927, took cover under the nearest available bed whenever he heard the distinctive "wa-peep-peep" of the blasting whistle during the 1930s. "It was altogether different than any other whistle out there," he said.

Storing, as well as setting off, dynamite could cause excitement and even danger. Ball remembered a time the "dynamite house" (possibly the magazine built in 1920) fell on its side. "It wasn't much bigger than a fish shanty," he said. "The dirt and gravel kept working out from under it and the blame thing tipped over one day. We had a stove in there, too. One of the fellows had nerve enough to run in there and grab the dynamite caps or the thing would blow up."

Shovel

Once a row of limestone had been blasted, a steam shovel would amble up to the blasted rock and load it into train cars that took it to the crusher. Early shovels were outfitted with railroad wheels, and workers laid and relaid track in ever-widening arcs to get them to rock that had been blasted.

Fiborn Limestone Company bought a Marion Model 92 shovel converted with caterpillar tracks in 1928 to replace a 95-ton Bucyrus ("Buckeye" to many workers) shovel, eliminating the need and expense for a crew to lay rails, and the cost of buying them. Like its predecessor, the Marion was powered by a coal-fired boiler and water pumped from the powerhouse. Workers laid and connected sections of pipe to bring the water from the powerhouse, running between the ties under railroad tracks and anywhere else needed. "We used to walk on those pipes all the time, us kids," Keith McEachern said. "My dad always cautioned me: 'You shouldn't be doing that, because you might break that pipe.'"

The quarry also had a steam shovel that stripped the thin layer of soil, brush, and trees covering much of the limestone. A shovel referred to in company paperwork as a "Bay City crane," which required four laborers to lay track, performed the task until the late 1920s, when the company bought an Erie B2 shovel with caterpillar tracks.

Steam-shovel operators were among the quarry's royalty. Tom Kelly ran the big shovels from 1920 or 1921 until possibly the quarry's

The Bucyrus, or "Buckeye" steam shovel, which ran on rails. The quarry's "big shovel" loaded blasted rock into train cars which took it to the crusher. (Aaron Thompson collection)

final years, and his importance to the quarry was reflected by the fact that Kelly and his wife, Blanche, both Indiana natives, lived in one of the block houses, even though they had no children, and they were mentioned often in the St. Ignace newspapers' Fiborn columns. "Tom Kelly with the buckskin belly," McEachern called him. "He ran the big shovel, so he was a big shot." Kelley was "real proud" of the Marion shovel, McEachern said. "He stood where all the controls were; all the levers, bouncing up and down on pedals.

57

The Buckeye loads blasted rock into gondola cars.
(Aaron Thompson collection)

A quarry worker stands next to the bucket of one of the steam shovels.
(Aaron Thompson collection)

The quarry also had a shovel that stripped soil from rock outcrops before crews drilled and blasted. (Aaron Thompson collection)

The 1920s version of the waste-rock shovel scoops fine rock from the waste pile. (Aaron Thompson collection)

And another guy rode right out on the boom. They had a little seat right out there and he just swung right with the boom. He handled the 'dipper stick.' So, if it was digging too hard, he'd ease her up a little bit." A third worker fired the boiler.

Another shovel carried fine waste rock from the sorting bins to pile it for later use as gravel, from a collection of photos taken by the family of Lynn Brockway, Fiborn's final superintendent, and preserved by his grandson, Duke Brockway.

Trains

Fiborn Quarry used two types of locomotives: one for hauling cars of rock to the crusher, another for hauling DSS&A cars filled with crushed rock to Fiborn Junction.

The engines used to haul rock to the crusher were smaller, and were commonly referred to as the "dinky" locomotives by quarry workers and village residents. They were more formally known as saddle-tank locomotives, due to the saddle-shaped water tank that sat over the boiler flues (long tubular spaces that carry heat). The dinkies pulled gondola cars whose bins tipped to unload their cargo. Clyde Ball was a "brakey" on one of the dinkies in the early 1930s. He said the job involved firing the locomotive's boiler, jumping off and running ahead of it to throw switches as needed, and pulling a lever on each gondola car to dump rock into the crusher.

David Beaudoin said his dad, Oliver, took him for rides to Fiborn Junction on a dinky to pick up meat for the store. "It was just the engine and no cars. We backed all the way back." Letha Shoemaker Derusha said her father, Bert Shoemaker, took her for rides early in the evening, when he helped load locomotives with coal and water for the next day's work.

The larger locomotives hauled cars loaded with crushed rock to the junction, where they were transferred to trains headed for Sault Ste.

An engineer and "brakey" on a "dinky," or more properly, saddle-tank locomotive,
used to haul gondola cars filled with broken rock to the crusher.
(Aaron Thompson collection)

Marie. Like the dinkies, the large locomotives were filled with coal and water each night, near the crusher. Water there was available only at night; locomotives that had to water up during the day when rock was being crushed used a large wooden tank located along the rail spur to the junction. Other maintenance included cleaning the fireboxes and flues.

Engineers such as Harry Myers, Lynn Brockway, Sid Bush, and Angelo Perone were second in importance to quarry operations only to the king

A saddle-tank locomotive hauls rock along the edge of the quarry pit.
The rails used were moved as the pit grew.
(Aaron Thompson collection)

of the big steam shovel. But with the prestige came risk. Bush's daughter, Anna Bush Anger, recalled an accident in which his legs were burned after a locomotive tipped over into deep snow and heat from the locomotive

The crusher (center, partially obscured)
being used in 1924 was possibly the "Mammoth McCully"
installed by S.B. Martin Company as early as 1910.
(Aaron Thompson collection)

fueled jets of steam from the snow. "They wanted my dad to crawl back into the hole," she said, "underneath the train. He said, 'No way, I was under there once and almost got killed.'"

Crush and Sort

The McCully crusher S.B. Martin Company installed in 1910 could process 1,500 tons of rock a day. Whether it was ever replaced, its capacity might have been enough for the quarry's lifetime of rising and falling production.

In such "Bell" crushers, a steam engine drives a large rotating pulley, which in turn drives a large "eccentric" (off-center) pulley with a belt. The eccentric pulley turns a large bell structure inside a larger, upside-down bell structure, producing an effect Keith McEachern described as "it rotated and oscillated at the same time." The rotating bell crushes rock up against the sides of the machine and the crushed rock falls to a conveyor belt, which carries it off for sorting.

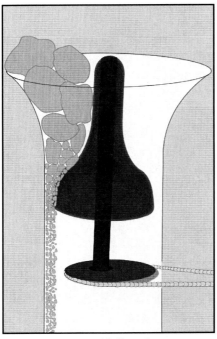

A typical bell crusher.
A metal bell structure mounted on a shaft running through an eccentric pulley is driven by a belt, or in the case of Fiborn Quarry a rope, encased in an inverted bell-shaped housing. The pulley drives the shaft in an off-center rotation that pushes the bell close to the walls, crushing rock caught between.

Don Stokes said workers occasionally had to splice the crusher's rope belt together after it would break. He recalled one particular worker, Rose Macco, splicing so quickly "you couldn't see his hands move."[49]

Fiborn Quarry's conveyors carried crushed rock up at roughly 45 degrees into a building where rotating screens of different mesh sizes shook through rock of certain sizes which then fell into loading bins from which it would be dropped into train cars. As engineers backed ore cars under the bins, a worker would sound a steam whistle one or more times to send messages such as "move forward," before opening the sorting-bin doors to load individual cars.

The top of the bell crusher pokes up from its housing.
(Aaron Thompson collection)

Danger lurked around the crusher, like much of the quarry equipment. Stokes recalled a fellow laborer accidentally being vaulted onto rocks in the crusher by a long metal rod he was using to pry loose rocks that jammed. Stokes said he managed to "pull him right out" after snagging the back of his bib overalls with a metal hook.

Powerhouse

The powerhouse contained a large coal-burning boiler that powered the crusher and helped drive the big steam shovel, and, after 1921, a 10-kilowatt electrical generator.

A machine shop also was crucial to quarry operations. The master mechanic was responsible for machining a wide range of parts to keep anything from trains to boilers working. "Those guys there, they were real machinists," Keith McEachern said. "My dad needed a part for the car and he didn't get satisfaction in the service garage, he'd say, 'I can get Matt Fillian to take care of that. He'll make a part better than General Motors.'" Emmanuel "Matt" Fillian was the quarry's master machinist in the mid-1920s to early 1930s. He was from Escanaba, of French-Canadian descent and lived in a block house with his wife, Florence, and two sons.

As workers fueled and watered locomotives or maintained the boiler overnight, a pair of night watchmen patrolled. One was stationed inside the powerhouse to keep an eye on boiler pressure and other things; another walked around the quarry, checking drills, steam shovels, and other equipment.

*Train cars were spotted under the sorting bins and rock from
the crusher fell through sorting screens into them.
(Aaron Thompson collection)*

In Case of Emergency ...

Accidents were a fact of life. When they happened, steam whistles sounded the alarm, sometimes in the middle of the night. "They'd blow the emergency whistle," McEachern said. "When they did that, everybody went running right to the plant. Or, if it was a steam shovel tooting, they ran right to the steam shovel."

Two views of the power house. The black pipe running from the front (foregound in the top photo, at left in the lower photo) carried water to the steam shvel that loaded rock into train cars.
(Aaron Thompson collection)

One middle-of-the-night alarm sounded after a watchman put his foot where it shouldn't have been while checking the boiler. "They had the inspection plate off, and this one guy went up looking at it and he put his foot in there. It just took the bottom of his foot right off."

McEachern and some friends witnessed the collision of two dinky locomotives one day while playing on a decommissioned steam shovel. One dinky, on a special job, was on the same track headed at a regular rock-hauling dinky. "These two trains tooted in unison and neither guy heard the other one," he said. "We're waving our arms, and they're waving

*The machine shop crew, as Aaron Thompson listed them: Blacksmith
and helper, boiler maker, and master mechanic.
(Aaron Thompson collection)*

back at us. They just ran right into each other." Luckily, "they didn't
get hurt that badly." Keith also saw a crew accidentally send a full-size
locomotive down a narrow track meant for dinkies, after which it took
two or three days to extract the locomotive and repair the track. He also
watched a crane topple while being moved by train after workers didn't
properly connect the car to a locomotive. The coupling slipped, sending
the crane car rolling freely. The crane's boom struck the sorting bins,
knocking the crane over. The crane operator jumped before the crane fell,
breaking his leg, McEachern said.

Living at the Quarry

The quarry made an intimidating first impression on visitors coming up the corduroy road from Caffey. "You'd come in that quarry road there, and all you could see was smoke and steam," said Riley Derusha, who visited the area in the 1930s before marrying Letha Shoemaker. She, however, remembered a peaceful, comfortable place. "I felt like I was living in a private little world. I was afraid to get out of it."

The occasional roar (and flying rocks) of blasting, the constant rumbling of trains and steam shovels, and the shrieks of steam whistles during the day gave way to the sounds of nature at night. "There used to be a lot of coyotes around there," Letha said. "We'd stand out on our porch at night and we would listen to those things across the pit."

Workers trudged back and forth between the boarding house and quarry pit for work, meals, and sleep. Families cut wood and scrounged coal to burn for heat, fetched water by the bucket, did laundry with washboards and buckets, and stretched their food budgets by gardening, canning, and hunting. Social life included Saturday-night dances and house parties, Sunday church services, and meetings of civic clubs. Children and adults played baseball in summer, skated on the frozen, flooded quarry floor in winter, and traveled by train to Newberry, Sault Ste. Marie, or St. Ignace for the occasional movie and shopping trip.

Villagers even celebrated Halloween with trick-or-treating. "We'd always bake cookies and give them to the kids, [also] apples," said Rose Stokes, who married Don and moved to Fiborn Quarry in 1934. "Never had much candy," she said, "unless you made it."

Though physical danger was an obvious, often daily, concern of quarry work, as well as life in an isolated, wilderness community, crime was rarely mentioned in surviving accounts of life at Fiborn Quarry. But it did occur. A *Republican-News* column in August, 1917, reported that a blacksmith at the quarry was beaten and robbed of $40, and a gold watch, while walking from Rexton to Fiborn Quarry. Nearly two years later, the paper reported that a burglar stole $70 from the Fiborn Quarry post office.

Few village residents owned automobiles until the late 1920s, even though a few turned up at Fiborn Quarry as early as 1914. Workers making less than $20 a week could mostly only dream of buying anything costing $500 or more, plus a lack of roads and snowplowing made the train a more reliable form of transportation. "We just about declared a civic holiday the day the first snow plow came to town," recalled Lillian Brockway in a 1965 interview with *The Evening News*. "We could hear the roar of the plow engines and pretty soon they came around the bend. We all had the feeling that at last we were free."

Boarding House

Along with housing and feeding workers, the boarding house was often the hub of social life in the village, hosting dances, church services, and even traveling salesmen who might whip up a meal if they were selling cookware.

The boarding house was 80 feet long and 40 feet wide. Its basement held a furnace and coal. The first floor contained two bedrooms for boarding-house workers, a kitchen, dining room, a large social room, showers and baths, and toilets. Bedrooms filled the second floor, each with a double bed, dresser, and steam radiator for heat. The boarding house had indoor plumbing, and when the quarry plant was running, electricity. Ice was stored in a heavily insulated shack tucked into the quarry's north wall a short distance from the boarding house. (A company in St. Ignace harvested ice on Moran Bay in winter and shipped it by train to Fiborn Quarry.)

The boarding house, built by 1907, offered a place to sleep and three meals a day to single workers, and served as a social center for the village. (Aaron Thompson collection)

Quarry workers climbed a stairway out of the pit to reach the boarding house.
Some younger village residents also used the wide rails as slides.
(Aaron Thompson collection)

When workers walked through the front door after climbing a long stairway out of the quarry pit and walking a couple of hundred yards or so, they could go directly down a hallway to showers and bathrooms, or into the social and dining rooms. The social room featured a piano, several residents recalled, and a pool table.

Along with a place to sleep and socialize, the boarding house supplied three meals a day. Mary McGillvray Houghton, whose parents ran the boarding house in 1928, remembered helping out by waiting on tables, and meals served table-style, with large bowls from which workers helped themselves. Pie was a lunchtime staple. "You had to have pie at noon."

Lillian Brockway, recalling long winters at the quarry, said they were made a bit easier by dances held on Saturday nights at the boarding house, often featuring an itinerant fiddler from some nearby lumbering camp. Sleighs often took people to and from the dances in winter. Fiborn Quarry residents also went to dances in Rexton, Garnet, and Trout Lake, especially as cars became more common. Keith McEachern said he and one or two friends played at some of the Fiborn Quarry dances. "I played guitar and mouth organ, and we had another guy with a banjo, and another guy with a fiddle." Letha Shoemaker and eventual husband Riley Derusha also used to perform, he on guitar and both of them singing.

Saturday night dances gave way to Sunday morning church services. Traveling ministers would visit and hold services, presumably in the social room with the piano in use and the pool table not so much. A *Republican-News* Fiborn column in February, 1918, noted a Baptist preacher from St.

71

Ignace, Boone Stigall, "has been holding religious meetings of late at Fiborn Quarry. His conversions were said to be great." Stigall collected signatures on statements professing the signees would refrain from alcohol and support Prohibition, which was in the process of becoming the law of the land. Keith McEachern remembered services in the late 1920s and early 1930s as "just straight preaching and sing some hymns, and say a few prayers. Pretty conventional." Catholic residents of the quarry went to Trout Lake for mass, and Vera McEachern, a Presbyterian, traveled with them to play organ.

Family Housing

Quarry workers who didn't live in the boarding house either rented a block home for $10 a month or a cottage for $5 a month. A few families built their own homes. Laborer Bert Shoemaker, when hired in 1926, tore down a board house he had built in Rexton, and rebuilt it at Fiborn Quarry, then later moved it as the quarry grew.

The block homes, located along the road east of the quarry works and boarding house, had two bedrooms upstairs, living room and kitchen downstairs, and basement with coal furnaces.

Fiborn Limestone Company added cottages to the village in 1917 and 1924, work mentioned in the *Republican-News'* Fiborn column. The company also bought four cottages from the nearby Wilwin Lumber Company in 1929. Horse-drawn wagons carried the cottages in sections to Fiborn Quarry, where they were assembled, painted, and wired for

Five block homs housed the quarry's top employees, such as the
bookkeeper and foremen, and their families.
(Aaron Thompson collection)

Bernard, Lillian, and Lynn Brockway (left to right) sit in front of the fireplace of their block home. Lynn Brockway was Fiborn Quarry's final superintendent. (Duke Brockway collection)

electricity at a total cost of nearly $4,000.[50]

Residents of company housing could remain in their homes rent-free when the quarry was shut down, according to former residents, and if the breadwinners chose to go elsewhere for work, they were obligated to return immediately when the quarry resumed operating to keep their homes. "If you didn't go back when they called you, you had to move out, and move your furniture out," said Don Stokes. "But if you'd come back and work, [even] if they only worked a week, you hold the rights to your house." This likely helped foster the shift to a village dominated by families rather than single workers. As jobs disappeared as the Depression set in, those that came with housing likely were particularly precious.

Electricity was available for four hours or so in the evenings, when the quarry was in operation. "They'd flick those lights at night, and all the kids had to get to bed," said Keith McEachern. "The adults would light up a gas lamp and stay up longer."

"We had lights, but we weren't allowed to use any appliances," said Belle Bigelow, whose husband Frank was an engineer. "We had radios, but we had to use batteries on them." Belle said she and a friend would carry batteries 2 or 3 miles on a toboggan to Norton's Lumber Camp, where they could charge them.

The boarding house and block houses had indoor plumbing and running water when quarry operation would power it. Village residents also got water from two faucets fed by a well, one near the block homes and one near the cottages. Some families, living along the railroad spur, got water from a spring and a creek. Most residents used outhouses—or worse—and bathed in wash tubs with water heated by stoves or open fires. They did laundry with washboards and buckets.

Some village residents lived along the railroad spur in homes they built. Longtime quarry employee Otto Nelson, a Swedish immigrant, had a log home listed as worth $250 in the 1930 census, where he lived with his Norwegian wife, Mary, son Melvin, and daughter Lillian. He bought 120 acres after living as a boarder for several years, had cows and sold milk to village residents, who remembered Nelson cross-country skiing to and from work, and riding into the village on a hand-driven rail car with milk to sell.

Engineer Oliver Beaudoin built what one son called "a tarpaper shack" for his large family when he was hired in 1926 or 1927. Another engineer, Sid Bush, laid claim to an abandoned shack near the spur, and employed his young children in adding an addition during their first years at the quarry. "My dad cut the trees down then we put a chain around our waist, and he'd push and we'd pull the trees up," recalled Anna Bush Anger. She and her sister filled gaps between the logs with moss. "If that moss come out, my sister and I would have snow on our bed in the morning." Anna said she, her sister, brother, and father shared one room "We had like a little hammock in the corner and my sister and I slept in that. My dad and my brother slept in the bed." They cooked and heated the shack with an oil drum fashioned into a wood stove.

One worker, a dynamite specialist who had emigrated from Poland, even lived in a small one-room shack literally tucked into the side of a hill, Keith McEachern said. "As you went up to the boarding house, it was right off on the right side." This likely was the former hen house that had served as the first office in the quarry, which Fiborn Limestone Company records mentioned as being converted into a "workman's cottage."

Food

Village residents in the late 1920s and early 1930s bought food at the company store run by the McEacherns, grew gardens, picked wild berries, hunted game, and even raised chickens to feed themselves. A few even made maple syrup. "We always canned up lots of stuff, made dill pickles and pickled beans, pickled corn. We all had gardens," said Rose Stokes. Except for the boarding house, no refrigeration or ice was available. "We canned our meat, if we had extra meat," said Belle Bigelow. She and other residents also had root cellars.

For one or two families who lived near it, Big Cave served as a refrigerator. "We kept buttermilk and eggs, and home brew down there to keep it cold," recalled Ed Shoemaker. "My mother used to have a pail with a cover on, she put all our butter and everything in that, and she'd set that, she'd fasten that right into the edge of that cave," said his sister Letha. "The cold water would run on that and I'll tell you it was just like you took it out of the fridge."

Hunting was a major source of meat. "In the wintertime, if we wanted a deer, we'd go out and kill one," said Don Stokes. "You had to eat." McEachern said Fiborn Quarry residents "hunted year round." Some snared rabbits.

On occasion, they rode the South Shore train to Newberry or St. Ignace to stock up on food and other supplies. "We did most of our shopping at Newberry, because it was closer," said Lillian Brockway.

The company store was in its second or third incarnation in the late 1920s and early 1930s. One opened in 1907, when the post office was established, and a new one was built in 1917 closer to the family houses as the quarry grew.[51] Don Stokes said workers could charge against their paychecks in the store. "I felt sorry for some of the people that had big families, because they'd get in debt and couldn't get out; they'd take it right out of your check." The store also held the post office. Mail arrived at Fiborn Quarry on a small specialty train running up from the Caffey station 3 miles or so south of the quarry. A telephone also was available there. "The only phone was at the store," said Stokes. "You want to call the doctor, you had to go and wake up Don McEachern and go to the store and call the doctor" (which Stokes did on a July night in 1934 as Rose went into labor).

School

Fiborn Quarry school launched with 11 students in 1907, according to Hendricks Township school census figures, and closed in 1935 with a class of 18, with anywhere from 15 to 25 typical of a given year.[52] At least

*Fiborn's one-room schoolhouse, soon after construction, judging by the building
materials lying around, from a photograph anonymously donated
to the Michilimackinac History Society.*

12 different women taught at the school over the years, several of them
teaching for a year, moving on to other schools, then returning.[53]

A wood stove heated the one-room school. Teachers had to fire the
stove, sometimes cutting and hauling wood themselves, and prepare lunch
for the students along with teaching duties. Students shared firewood
chores, and drank from a bucket of water drawn from the nearby creek.
Teachers worried about the creek flooding in spring and warned their
young charges to stay away from it during recess.

"You could ride to school on a sled in the winter," recalled Norma
Bigelow Malardar, whose father, Melvin, worked at the quarry during the
late 1920s along with sons Frank and Earl.

Students attended Fiborn Quarry through seventh grade, and then
went to Rexton for grades eight through nine, and St. Ignace for grades
10 through 12. They had to pass state examinations to go from seventh to
eighth, then ninth and tenth grades. "In those days, each grade was a big
deal," said Keith McEachern. "If you passed the eighth grade that was like
coming out of high school."

The school also served as a health center, where traveling county
nurses occasionally performed checkups and gave talks on health and
hygiene.

Blasting, crushing, and hauling rock weren't the only hazardous
jobs at Fiborn Quarry. Even teaching school took a toll, beginning with
the tale of Miss Florence Cheeseman. The St. Ignace *Republican-News*
in September, 1912, reported she "has been engaged for another school

year at Fiborn Quarry, where her ability as a teacher is fully recognized, this being her third term. Though Fiborn Quarry is a little remote, Miss Cheeseman has the compensation of a very desirable position in a modern school building, complete equipment, an attractive salary, and a progressive school board. City people might wonder how rural communities like Fiborn Quarry manage for recreation, but in truth, they become more resourceful than townsfolk, and their pleasures are more rational, while their social relations are far less constrained and more enjoyable. The teacher who has a good country school is far better off than her city cousin in health, pocket, and recreation; and from a professional point of view she is usually, when once experienced, the better, because the more normally balanced, teacher. [*sic*]"

Those benevolent forces of country living seem to have turned against Miss Cheeseman, who wound up on the disabled list in just a few weeks. The *Republican-News'* November 26 edition reported that Florence, "who sprained her knee in the schoolroom about five weeks ago and was compelled to come home, was able to be out on crutches. During her incapacitation, the school has been closed." The injury forced Cheeseman to give up the Fiborn Quarry job. Despite or even because of the extended recess, Fiborn Quarry's school children, if not their parents, hopefully became more resourceful in pursuit of recreation, and their social relations turned far less constrained and more enjoyable.

Elta Norton LaCount, who taught from 1927 to 1930, proved a teacher in such a remote setting had better be hale and hearty. Elta, who lived at Norton lumbering camp, which her father owned, walked to and

Boys play tug-of-war near the schoolhouse.
(Emma Kalnbach collection)

*Teacher Elta Norton LaCount (right, standing)
and some of her students.*

from Fiborn Quarry, and snowshoed and skied in winter. "My dad made me a trail through the woods, so I went on showshoes," she said. "I was pretty good on skis, too."

After arriving at school, she would fire up the woodstove and often begin preparing a lunch which would simmer on the stove during morning lessons, then summon her students playing in the nearby woods by ringing a hand bell. Elta remembered bean soup and potato dishes as typical fare. She had a young son who she sometimes had to bring to class. The toddler was relegated to playing under her desk while she taught.

Elta stayed with the McEacherns overnight at times of bad weather early in her Fiborn Quarry tenure, and by 1930 she and her son moved in with them.

In May, 1928, she became ill in class, possibly suffering appendicitis. "Elta LaCount was brought to the local hospital May 1 suffering from a complication of ailments," reported the *St. Ignace Republican-News* on May 12. "She underwent an operation that same day and is now in a satisfactory condition." Elta said she was taken down the railroad spur to Fiborn Junction on a sled fashioned from the hood of a car or truck, pulled by horses.

Elta moved on, and Emma Kalnbach took the teaching job in 1931. Emma, like her predecessor, was remembered fondly by former students.

"There was no janitor, she was the janitor, and what the boys done for her," said Marjorie Peters Derusha, who lived at the quarry in the 1930s. "When they delivered wood, she'd split it and we'd carry it in and put it in the back of the school room." Emma also insisted on good hygiene. "Before school started, everybody had to put their hands on the desk," recalled Ed Shoemaker, Letha's brother. "She'd go along and turn them over, see if they had clean fingernails, and clean hands. ... If you went a whole week and didn't have dirty fingernails, you got a gold star at the end of the week."

Along with a blackboard and desks, the school had a very small library.

The school's last class, with teacher Emma Kalnbach (far right) in 1935.

The books included "the Bobsie Twins and stuff like that, Huckleberry Finn," Letha Shoemaker Derusha said. "She'd read us a story every day, too, before we'd start school."

Fiborn Quarry students participated in events with other schools, such as field days and spelling bees. "The Christmas programs that we used to have at our little school were something else," Letha said.

Recreation

When not in school, village children did many of the same things for recreation as kids everywhere, and some unique to an isolated, almost wilderness setting. Marjorie Peters Derusha said she and a brother used to walk 6 miles to Rexton on Saturdays to play, then walk back to Fiborn Quarry. "Leave in the morning and be late when I come home at night," she said. Keith McEachern and friends played in a sandpit, imitating their dads down in the quarry. They also used the railing of the stairway leading into the quarry pit as a slide.

In summer, some kids swam in a quarry water tank while others, particularly those who lived near the rail spur, swam in a creek (with one important *caveat*: Stay upstream from the pipe that dumped raw sewage from the quarry works and boarding house). They used scraps of wood and metal to dam up the creek and form a small pond. "We had a diving board out there and everything," said Keith Peters (Marjorie's brother). "Sometimes I think maybe we were dirtier when we came out than when we went in, but we loved to get in it anyway," said Letha Shoemaker Derusha.

In winter, the creeks and ponds, as well as water that frequently collected in the quarry, froze and allowed for skating. Villagers frequently built bonfires for night-time skating parties there. Kids also went sledding on hills near the boarding house and quarry, and one or two had skis. McEachern recalled skiing down a hill near the houses by the light of a lantern hanging from a tree. "In the wintertime we never missed one night of sliding downhill," said Letha, noting they mostly used pieces of cardboard.

Baseball was king of sports. Local newspapers reported on leagues formed across the region, sponsored by lumber camps, Civilian Conservation Corps camps, quarries like Hendricks and Fiborn, and villages. Don Stokes could legitimately claim semi-pro status. "[A business owner] gave me ten dollars a game to come and catch for Trout Lake," he said. Engineer Frank Bigelow, who also served on the Hendricks Township board in the early 1930s, was "quite a good player," said Belle. "We'd leave about eight o'clock in the morning to get to St. Ignace in time to play baseball in the afternoon," she recalled. Howard Alkire remembered playing on a team sponsored by quarry superintendent Harry Myers. "He bought all the kids baseball suits," Alkire said, and in return they cleaned a coal mess behind Myers' house. Villagers also regularly played baseball and softball on one diamond between the boarding house and quarry, or another closer to the family homes.

Some of the "big kids" also hunted for more than food. "A lot of those guys there belonged to the Trout Lake Rod and Gun Club," McEachern said. "They had a skeet shooting range there, and the whole works. ... In the fall, they'd choose up sides and have a game hunt contest." Tournament results turned up in the St. Ignace newspapers during the early 1930s. Ed Shoemaker trapped muskrats and coyotes, and recalled one of his first trapping expeditions ending with the capture of a skunk. Shoemaker checked his traps one morning before school, discovered the skunk, and proceeded to pick up plenty of its odor disposing of it. He then went to school, where teacher Emma Kalnbach promptly sent him home.

Last But Not Least, the Cave

Chase Osborn's "Big Cave" was located southeast of the quarry works, and remained until being mined away after the quarry closed. A remnant of that cave remains in the south wall of the quarry. Villagers who explored Big Cave described it much as State Geologist A.C. Lane did in 1901: a low, wide walkable passage still being cut by a stream, entered at one end through one of several small sinks, and at the other through a large sink containing a whirlpool where water sumped into deeper passages on its way to the Hendrie River. "They had a ladder going down into it, a railing around it," said Keith McEachern. "You could go one heck of a long distance."

"When I first moved into the quarry as a little girl, we were right close to what they called 'the deep cave,'" said Letha Shoemaker Derusha, whose father rebuilt their Rexton shack close to the quarry's south wall in 1926 or 1927. Letha said there were other caves near their house, "but they were kind of shallow." The Shoemakers were among those who stored butter, lard, and even home-brewed beer in the cave's cold, flowing water. "In the spring that cave would fill almost to the top with water. They had a fence around it. ... And my mother lived in constant fear that one of the kids was going to fall in that cave."

Letha's mother, Velma, wasn't just worrying in vain. Brothers Edward and Leonard decided one day to explore far beyond the pit entrance, with its ladder and open passage. Letha and sister Dora followed. "We went and we took a lantern," said Letha. "Talk about scared. We got

*Byron Brockway stands by the entrance to the Big Cave during a visit
to Lynn and his family, possibly in 1938. This shows the cave
was quarried out after Fiborn Quarry officially closed. Road
commissions and private contractors often struck deals to quarry
rock for road-building and other purposes in the years after Algoma
Steel Corporation shut down Fiborn Limestone Company.
(Duke Brockway collection)*

down there and it was so dark other than the light from that lantern.
Water was dripping off the rocks over our head. ... We walked in water all
the way." Finally, Letha and Dora saw daylight and climbed up through
an opening in the rock. Ed and Leonard, however, pressed on. "It should
have taken us only about 10, 15 minutes, to go through there," Ed said,

"but half an hour goes by and we're not out of there yet. And Letha goes and tells my mother that we're down there in that cave." The brothers finally crawled out through a small sinkhole not far from where Letha and Dora emerged, to meet a mother "madder than an old wet hen."

"Some places we had to crawl," Ed recalled. "It was quite wide, but a lot of spaces were just a couple feet [high]. And then there were other places where you could get up and stand up and walk."

The End

Like any manufacturing business, especially one reshaped by a bankruptcy and receivership, Algoma Steel wanted raw materials at the lowest price possible. Fiborn Quarry, with its dependence on rail shipping, was ultimately doomed by the rise of more cost-effective marine shipping and the availability of high-quality limestone from the growing Michigan Limestone operations in Rogers City.

The cost of maintaining the 30-year-old quarry's equipment may have been a major factor in Algoma Steel's decision. After a brief reopening late in 1935, Algoma shut down the quarry at the end of January, 1936. The post office was decommissioned on January 31. Newspapers in Sault Ste. Marie, St. Ignace, and Newberry carried no mention of the quarry's last attempted startup or its final closure.

The Marion steam shovel was taken out on hand-laid rails and sent by railroad to another quarry. Machines were scrapped and sold to salvage firms. The crusher was removed and the sorting bins torn down. Keith McEachern remembered seeing quarry workers cut up the locomotives into scrap to be recycled into new steel by the blast furnaces at Algoma. "That must have been traumatic for those guys. Here you are cutting up what was your livelihood, and what have you got ahead of you?"

Lynn Brockway stayed on to oversee the quarry's dismantling and the property. He dealt with road commissions and private contractors who brought in their own crushers and hauled away rock for road-building and other projects. Brockway and his sons, Harry and Bernard, also fished, hunted, and trapped the surrounding woods. Bernard often travelled those woods on a horse he owned for several years.

The boarding house burned down in 1936, said Bernard, who recalled running up to the burning building with his father and finding it totally engulfed in flames. No cause was ever determined, he said.

McEachern recalled taking a friend to the former quarry after graduating high school in 1937, bicycling from St. Ignace. "A buddy and I went there and stayed in [our] house," he said. "It was still there then with all our furniture in it. ... Everything was gone then except the houses."[54]

Frank and Belle Bigelow bought their old block house and moved it a few miles to Caffey Corner. "We had a tavern and a gas station ...

Lynn Brockway takes a break while snowshoeing in the quarry
pit. After leaving Fiborn Quarry, Brockway managed
Ozark Quarry near Trout Lake.
(Duke Brockway collection)

lived there about 10 years," she said. They sold it in 1941, when the new, rerouted highway US-2 opened farther south along the Lake Michigan shore, and the Hiawatha Trail became much less travelled. Cottages were sold and moved, turning up in Rexton and Caffey.

Mackinac County's first aviatrix, Lillian Nelson, made front-page headlines by landing near the quarry in 1937 to visit her parents, Otto and Mary, still living in their log cabin.

The block houses were torn down in the 1940s, Bernard Brockway said, and about that time Lynn and family left Fiborn Quarry for Trout Lake, where he managed the nearby Ozark Quarry until it closed in

1945.[55]

For years, people living in the region often procured a pickup-truck full of gravel with no notice by the land's absentee owners. Many searched the area for bottles, pieces of machinery, and other relics for years after the quarry closed. The river cave, later known as Hendrie River Water Cave, was well-known among locals, and attracted a steady stream of visitors.

Fiborn Limestone Company existed on paper for 30 years after the quarry closed, owning Northwestern Leather Company in Sault Ste. Marie and other properties for Algoma Steel. Fiborn Limestone Company was finally dissolved by circuit court order in 1965, according to *The Evening News*.

In 1987, Algoma Steel sold 480 acres of the quarry property to the Michigan Karst Conservancy, which created Fiborn Karst Preserve to preserve the land's natural and historic features. The preserve is open to the public, with limits and conditions intended to protect its natural

Bernard Brockway on his horse. Bernard and his brother, Harry, fished, hunted, and trapped the woods and swamps surrounding Fiborn Quarry for several years while their father, Lynn, oversaw the quarry's dismantling, negotiated with contractors seeking crushed rock, and otherwise supervised the property. (Duke Brockway collection)

features. The conservancy holds regular work weekends at the preserve from May through October, and during those weekends guided cave trips are offered by experienced members of the conservancy, who also are members of the National Speleological Society. The preserve also has foot trails highlighting the area's natural and historic features.

Now: Management of the Fiborn Karst Preserve

Michigan cave explorers became aware of the caves of Fiborn Quarry due to an article in the September 1, 1901, issue of *Michigan Miner* magazine. Several cavers spent long weekends in the Upper Penninsula searching for those caves. In the 1970s Bill Fritz discovered the location of Hendrie River Water Cave in a conversation with local people in a bar in Trout Lake. Bill says the information cost him a case of Budweiser.

On November 17, 1983, a group of concerned individuals, mostly cave explorers, established the Michigan Karst Conservancy (MKC). The application for a charter was signed by Mark A. Navarre, Rane L. Curl, David Luckins, William G. Fritz, Alice Rolfes Curl, and Virginia G. Yates on November 2, 1983. The purposes of the MKC are acquisition, management, and protection of karst features and areas; scientific study; and conservation education regarding karst in Michigan.

In 1987, members of the MKC became aware that the Fiborn property, including Hendrie River Water Cave, was being offered for sale in 40-acre tracts. To the MKC, this was an emergency. After several meetings where donations were collected and substantial loans were negotiated from two members, the 480-acre Fiborn Karst Preserve was purchased by the MKC that year. In 1988 The Fiborn Karst Preserve was registered with the Michigan Natural Area Registry, a program of The Nature Conservancy to identify and help protect significant natural lands and features in Michigan. The Preserve was the first site included in the Registry primarily for its unique geological features.

The preserve is managed by the MKC's Fiborn Karst Preserve Committee. The Conservancy manages the preserve as a natural area, open to the public under guidelines meant to prevent damage to natural features; vandalism; and unsupervised, unsafe cave exploration. MKC volunteers gather once a month from May through October to perform maintenance and management work such as clearing and mowing trails, posting or repairing signs, and showing visitors the history, geology, and biodiversity of the area.

The first project in the new preserve was establishment of a parking area. Bill Fritz built a six-sided history pavilion at home and members assembled it near the parking lot. The Fiborn Quarry History Pavilion was dedicated on August 15, 1992. Several members constructed an outhouse nearby for use of visitors The History Project provided documents and photographs illustrating the history of the quarry to add to the pavilion. It was later rededicated in 2012 as the Emma Kalnbach Pavilion in honor of the last school teacher at Fiborn Quarry.

MKC member Paul Johnson stands beside the Emma Kalnbach Pavilion.
Photo by Bill Greenwald, June 21, 2014.

Over 2 miles of self-guided trails offer visitors a look at the geology and history of the area. The trails loop past shallow sinkholes and disappearing creeks, wind through second-growth forest, and showcase the remnants of the buildings that supported the quarry operations. Historical features include the concrete portion of the machine shop and the remnants of the power house and the sorting and loading building in the heart of the old quarry. Guided tours through Hendrie River Water Cave are available upon request.

Protection of natural features and the safety of visitors are major concerns. For these reasons managing policies have been adopted for the use of the Fiborn Karst Preserve.

The following are not permitted in the preserve: fires, camping, littering or dumping, altering any natural features, collecting firewood, vehicle use off-road, using firearms except during firearms-hunting season, trapping, or polluting soils or streams.

Permission from the Fiborn Karst Preserve Committee is required for climbing or exploring caves, scientific collecting (of minerals, flora, or fauna), excavating or moving soil or rocks, and installing scientific instruments. Climbing and cave exploration in the preserve are permitted only if an experienced climber or caver (as determined by the Preserve Committee) accompanies each group, and a liability release and acknowledgement of the Preserve Use Policies is signed by every visitor.

The Quarry Works

A few remnants of the quarry works remain today and are considered important historic buildings. As recounted in the last chapter, the boarding house burned down in 1936. The block homes were torn down in 1940. A subsequent owner filled in or destroyed the foundations of the block homes and there is almost no trace of them today.

The foundation of the boarding house seen from the rear today.
Compare page 70. Photo by Mark Whitney, May 18, 2014.

Every scrap of metal—pipes, machines, rails—has been salvaged long ago. The wooden part of the machine shop has disappeared but the concrete portion remains. It has suffered serious vandalism in the past. The Michigan Karst Conservancy is making an effort to preserve what

The foundations of the block homees can be found only by diligent searching.
Photo by Bill Greenwald, June 21, 2014.

The sorting bins and loading ports today.
Compare page 65. Photo by Bill Greenwald, June 21, 1014

The power house today.
Compare page 66, lower picture. Photo by Bill Greenwald, June 21, 2014

The machine shop today.
Photo by Tim Deady, October 20, 2012.

The quarry works today.
Compare page 53. Photo by Bill Greenwald, June 21, 2014

remains. Concrete portions of the sorting bins and the power house remain to be seen.

The Caves

Hendrie River Water Cave exists today much as it did during the quarry days. The cave was explored and mapped in 1975 and 1976. The air-filled portion of the cave extends for about 1,800 feet where the passage descends below the water table and continues as an unenterable, water-filled passage for about a mile to where the water emerges as a spring and flows a short distance to the Hendrie River.

This cave is dangerous for unexperienced cave explorers. It is cold and wet. A disabling accident could result in death from hypothermia as the cave is deep in the forest and no rapid rescue will be available. Written permission is required from the Michigan Karst Conservancy to visit the cave. Entering the cave without permission is trespassing and trespassers, if apprehended, will be prosecuted.

The spring resurgence was dived in 1988. A condensed version of the divers' report is included below:

On Saturday, December 3, 1988, Bruce Herr, Saginaw,

The Bridges Entrance to Hendrie River Water Cave.
Photo by Tom Rea, about 2004.

Michigan; Dale Purchase, Saginaw, Michigan; Steve Omeroid, Maryville, Ohio; and Mark Pansing, Columbus, Ohio, participated in a dive at the resurgence.

Bruce and Dale had entered the resurgence several years earlier shortly after it was located and partially excavated by

The Resurgence.
Photo by Tom Rea, about 2004.

95

the Michigan Interlakes Grotto, a chapter of the National Speleological Society. At that time they determined that there was enterable passage. Using a 100-foot-long hose connected to compressed air on the surface they were able to examine a low, wide passage that continued on with one branch to the left.

The new dive, with a penetration of around 250 feet, uncovered more details of the types of passages and several additional going leads. All of this is seen as eye witness evidence of a substantial cave system that was only previously supposed by surface analysis and dye tracing.

Dale Purchase was first in the water. He examined the entrance, which inclines down 45 degrees for two body lengths to the floor of the passage. Dale dived the spring and had to adjust his tank, finally pushing it in front as he made his way slowly down. His fins were visable on the surface as he fought to move some rocks to the side and ease downward. Mark dived next. He proceeded down the entrance and out of sight. He reported a penetration of 100 feet.

Bruce would swim lead on the final dive. Mark followed Bruce along the dive line as Bruce played it out. In the next 10 or 15 minutes the cave finally yielded to the hours of careful planning and assault.

Five or six passages are easily enterable in the first 200 feet. The passages are mostly of a standard size. The flat ceiling entrance passage gives way to an area of more or less uniform low canyon passage with 3 or 4 feet of clearance. Passages are not well defined. The cave appears to be going in many directions.

The ceiling and floor do not meet. The width of the cave is dissolved on 4- to 8-inch bands on a bedding plane, giving the impression that nothing is supporting the ceiling.

No branches, leaves, or sticks were visible in the silt. This is consistent with the fact that all known insurgences are thousands of feet distant. The sump in Hendrie River Water Cave is about a mile away.[1]

The "several small sinks" which State Geologist A.C. Lane described on page 19 as the south end of Big Cave, exist today in the woods south of the quarry pit. The majority of Big Cave no longer exists. The water that

1   Purchase, Dale, "NSS CDS Dive Team Penetrates Fiborn Resurgence," *The Spelean Spotlight*, Vol 17 No 12. Reprinted in *1988 Speleo Digest*, pp 96–98

The southern sinkhole entrances to Big Cave.
Water flows underground from one to the next.
Photos by Bill Greenwald, June 21, 2014.

flowed through the cave now exits through blasting rubble in the south quarry wall and flows across the surface to the sinkhole described by Lane as 30 feet deep. It is now about 10 feet deep and was recently dug to the water table searching for additional cave passages.

Two other caves exist on the Preserve. The second longest cave is Kochab Cave. This cave has 485 feet of low, wet, muddy passage. In one location the ceiling is quite unstable and is in danger of collapse. This cave cannot be visited. Disgusting Cave is. It has approximately 100 feet of low, almost impassable, passage which is important as another drainage path from the wetland to the karst network.

97

The water flowing in the remains of Big Cave exits the quarry wall through blasting rubble. It is not possible to enter the cave here.
Photo by Bill Greenwald, June 21, 2014.

There are two other sinkholes in the floor of the quarry located near the machine shop. During wet weather these sinkholes swallow tremendous amounts of water and transfer it to the resurgence. It appears that that one resurgence handles all of the water that sinks into the ground throughout the quarry.

The Trails

Two hiking trails have been established on the preserve to demonstrate to visitors the karst nature of the area and to exhibit artifacts of the quarrying operation. Both trails start at the trail head across the road from the parking lot and pavilion. Printed trail guides for each trail are available near the start. Both trails return to the parking area. The trails are rough in some places and are not suitable for wheeled vehicles.

Dr. Rane L. Curl Sinkhole Trail

The Sinkhole Trail is a short loop trail starting at the trail head across the road from the east end of the parking area. Proceed south from the Trails sign, until you see the post marked 1, turn west (right) and follow

The "whirlpool" that still carrries away the water from Big Cave.
It has been dug to the water table to investigate the cave passage.
The automobile is on the route of the railroad track seen in the picture on page 82.
Photo by Bill Greenwald, June 21, 2014.

the trail and marked posts from there. The trail will return to Norton Camp Road about 200 yards west of the parking area. If at any time you lose the trail, you can return to Norton Camp Road by heading due north through the forest.

The trail goes approximately west along the northern margin of an area in which many sinkhole features have developed in the Fiborn Limestone bedrock. We believe that the swamp now located more to the south once encroached upon the higher ground in this area, probably assisted by the ponding of the swamp by beaver dams. The water began to dissolve limestone in joints in the bedrock to find its way underground to the South Fork of the Hendrie River, about a mile to the northeast.

The upland forest throughout the Preserve, which you will pass along this trail, is a second-growth beech–maple hardwood forest, consisting of sugar maple with scattered trees of beech, hemlock, balsam fir, and white cedar in the lower, wetter areas. Red-berried elder is common, while the ground cover consists of woodland fern, northern wood sorrel, and ground pine. Red raspberry is common in open areas.

The development of the sinkholes has occurred during the past 8,000 to 9,000 years, after the glacial Lake Algonquin that covered this area receded. The relatively great speed of the development of the sinkholes may have been due partly to the acidity of the swamp water that ponded upon the limestone, since limestone dissolves more readily in acid

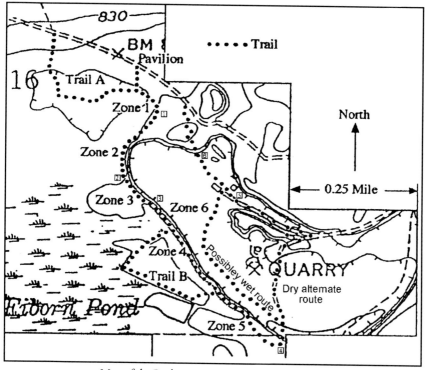

Map of the Barbatra Ann Patrie Memorial Trail.

100

solutions. The limestone bedrock here has been dissolved to some depth so it is buried in moderately deep sand and is not exposed.

Barbara Ann Patrie Memorial Trail

The Barbara Ann Patrie Memorial Trail is a 1.5-mile loop beginning and ending at the east end of the parking area. Walk south from the Trails sign to where both the Sinkhole Trail and the Barb Patrie Trail begin. Turn left to the Barb Patrie Trail. The large boulder with the commemorative plaque is an intrusive (plutonic) rock called gabbro, which was transported from the Canadian Shield (pre-cambrian rocks) by glaciers more than 9,000 years ago. Figure 1 shows a portion of the Preserve and the Barb Patrie Trail.

Bedrock is exposed at a point where overlying soil (overburden) had been removed between the trail and the quarry. Good examples of karst pavement, open joints, and solution-widened joints are visible here. These formed below the original soil cover. One may cautiously descend a narrow defile to the quarry floor near here to inspect a section of the pure Fiborn Limestone exposed by the quarry. The characteristic low-angle joints of the formation are seen here. The quarry floor is a magnesian limestone that was not suitable for the uses intended for Fiborn Limestone.

A junction with the Beaver Pond Loop is encountered. Beaver Pond Loop climbs up onto the overburden dumping area that abuts Fiborn Pond and eventually circles back to the Barb Patrie Trail. From its northwest (first encountered) end, the Beaver Pond Loop ascends along the northwest end of the overburden dump. The very pronounced and distinctive ridges were formed by overburden being dumped to both sides from rail cars before the tracks were moved over to form the next pile. The trail ascends to an overlook into Fiborn Pond. Originally the pond was a cedar swamp and the stumps left from logging the cedars are still visible. Beavers live in and maintain the pond. Their trails and aspen cuttings are visible along the loop. The southeastern dam is visible where the loop turns to descend along the edge of the woods back to rejoin the Barb Patrie Trail.

Barbara Ann Patrie

Barbara Ann Patrie was the first trip coordinator and most active trip leader for the Fiborn Karst Preserve after its creation. She shared her enthusiasm for this Preserve and the work of the MKC with all she guided here. She helped begin the Fiborn History Project oral history program. She usually started her tours at the southern overlook, where Sinking Creek flows into the quarry, so visitors could learn about the history of the quarry and ghost town of Fiborn Quarry as well as about karst and the natural features of the Preserve. She served on the MKC Board of Trustees from 1988 until her untimely death in 1991.

Geological Setting

The Fiborn Limestone:
Geological History, Quarry Area Features, and its Place in 19th Century Sault Ste. Marie Industry

Rane L. Curl
Professor Emeritus, University of Michigan

Michigan's Upper Peninsula in the late 19th century was attracting people and industries because of an abundance of natural resources, including furs and forests, iron and copper ores and other minerals, water transportation, and hydroelectric power. Availability of dolomite was well known, but discovery of limestone spurred interest in developing additional industries that required it, such as for steel, paper, and chemicals manufacture. This chapter describes some of these events and developments, which were important in both quarrying at Fiborn and its eventual demise.

The book *Geology of Michigan* (Dorr and Eschman, 1972) describes the principal geologic features of Michigan and their origins from the earliest events recorded in the rocks to the present day, and is recommended for placing the Fiborn Quarry story within the broader sweep of geological history.

Geological History

When one drives south on I-75 from Sault Ste. Marie, over an area of low relief, a range of hills appears ahead. These are of the Niagara Escarpment, eroded remnants of a sequence of resistant sedimentary layers of primarily carbonate rocks (limestone, dolostone, and dolomite), which rim the Michigan Basin. The escarpment forms Niagara Falls and cliffs and hills through the Bruce Peninsula in Ontario and across Michigan's Upper Peninsula, and southward into the Garden Peninsula and Wisconsin's Door Peninsula. Once one drives onto the escarpment the land has generally low relief though with some local bluffs and river channel, until near Lakes Huron and Michigan, having been planed that way by southward moving glaciers of the Wisconsin (and earlier) ice ages. The rock strata slope (dip) south at only 40 to 50 feet per mile so the drive south crosses a succession of younger aged rocks. The total vertical thickness of Niagara Escarpment rocks (the Niagaran Series) is some 550 to 800 feet, depending on where one is along the Escarpment.

Rocks of the Niagaran Series were formed from sediments consisting primarily of skeletal remains of marine organisms, ranging from bacteria and algae to a wide range of invertebrate marine animals including corals and a long-extinct group known as stromataporoids (related to sponges). These latter two groups were reef builders, and the Niagaran Series is famous for fossil reefs. In addition to the skeletal remains of fossils, carbonate muds also accumulated in abundance, eventually forming rocks that include some high calcium limestone (mainly the mineral calcite, a calcium carbonate, $CaCO_3$), much dolostone (magnesian limestone containing magnesium carbonate, $MgCO_3$), and mostly dolomite, $CaMg(CO_3)_2$. These were deposited over a span of about 21 million years during the Silurian period, about 430 million years ago, when warm and shallow tropical seas covered a large depressed portion of the central North American continental plate while it was located just below the equator, and connected by channels to the much deeper surrounding ocean (Shaver, R.H., 1996). Rock layers of this geological system underlie the Michigan Basin like nested "saucers," due to slow sinking of the central areas of Michigan over millennia, such that they lie nearly 5 miles below the surface in mid-Michigan, where they are overlain by younger rocks.

Originally these rocks extended over a much larger area north of the current Escarpment, well into Canada as far as Hudson Bay, but have since been eroded back by glacial scouring, rivers, and chemical weathering. Glaciers also transported and deposited thick layers of glacial drift—sand, clay, and gravel, derived from granitic rocks of the Canadian Shield. Glacial drift covers subdued portions of the Niagara Escarpment and underlies the extensive swamps and forests of the Upper Peninsula.

During marine deposition of these Niagaran sediments water depths often changed as sea level rose and fell, and sediment compositions changed as the chemistry of the sea changed or energy conditions changed. Occasionally sea level fell exposing the deposits to weathering and erosion. Runoff from land areas far from the basin, or current-swept sediment off shores and shoals, brought in sand and clay to mix with the carbonate materials, resulting in many rock layers of different compositions and textures. Evaporation of water from covering seas concentrated minerals that included magnesium salts, which reacted with limestone to partly or completely convert it to dolomite $(CaMg(CO_3)_2)$. Conversion to dolomite occurred to most of the original limestone of the Niagara Escarpment after its deposition, but a few layers escaped conversion, including most notably for this narrative, the Fiborn Limestone.

Rock layers of the Niagaran Series are divided into several *formations* and *member units*, based upon their composition and fossil assemblages (Figure 1).

Formation	Thickness (feet)
Engadine dolomite	200 to 250
Cordell dolomite	135 to 150
Schoolcraft dolomite	40 to 60
Hendricks dolomite Fiborn limestone member	60 to 120 (18 to 50)
Byron dolomite	80 to 155
Lime Island dolomite	15 to 35

Thicknesses of rock formations in the Niagaran Series (from Ehlers and Kesling, 1957). Fiborn limestone is a member unit in the upper part of the Hendricks dolomite formation.

(The naming of formations may change with further study of the rocks. Since Ehlers and Kesling (1957) the Engadine dolomite has been designated a group and divided into three formations. Catacosinos *et al* (2000) created a recent chart of Michigan rock strata names.)

Formations are distinguished from one another by major changes in the character of the rock and by their fossil content; contacts between formations may represent major erosion events, sea level changes, climate changes, or changes in the relative abundance of different fossil species. The names of formations are usually derived from place names where geologists first sampled and described them. Even within each formation there are layers, formally *members* (or *units*), that differ in smaller but similar ways as between formations. Ehlers and Kesling (1957) distinguish 15 such units within the Hendricks Formation.

It is impossible to describe all the events that effected the deposition of the rocks of the Niagara Escarpment so many millions of years ago. There were long term climate changes due to earth orbital shifts, seasons, tides, sea level changes, and currents carrying the marine waters of the surrounding oceans into the basin and water of the continental seas back

105

to the oceans. Shaver (1996) provides an introduction and summary of studies of these processes. Distant volcanic eruptions produced ash, carried by winds and currents to form thin layers among the carbonates. The average rate of deposition of up to 800 feet of carbonate rocks in some 21 million years is less than the thickness of a sheet of paper per year.

What with sea level changes, daily tides and storm, the growth and deaths of generations of organisms in shallow and deep water, with waves and currents grinding parts of the fossil material to produce carbonate muds—and all this varying across the area—it is marvelous that geologists have been able to put together even a rough description of what appears to have happened over those 20 million years.

For example, mud cracks, and suspected worm tracks, in the unit just below the Fiborn Limestone in the nearby Hendricks Quarry, indicate a period of exposure and drying of the carbonate mud before the seas returned and deposition continued. A mystery is why the rather thin bed of Fiborn Limestone, after deposition, was not also converted to dolomite like most other limestones were. Rocks of the same age east of Chippawa County and west of Schoolcraft County are almost all dolomites. Perhaps it was because the Fiborn Limestone was formed from a very finely ground fossil carbonate "milk" that settled as mud and consolidated in deep, quiet water, so that it became too compact for later sea water, concentrated in magnesium by evaporation, to infiltrate it and react to convert it partly or totally to dolomite.

Quarry Features

Of particular interest to geologists and *entrepreneurs* of the late nineteenth century were beds of moderately pure ("high calcium") limestone, such as Fiborn Limestone, which could be used directly in the smelting of iron and in the manufacture of cement, and to make lime (calcium oxide, CaO, also called quicklime) for use as plaster and in sugar refining, paper making, and other applications. Fiborn Quarry is the *type locality* (where the geologists who named it first studied samples) for the Fiborn Limestone, but the unit was also exposed at the surface at nearby locations, where the Blaney (Nicholsonville) and Hendricks quarries were developed, although the fact that these were all the same limestone unit took some time to establish.

The three quarrying sites at Fiborn, Hendricks, and Blaney were developed first because Fiborn Limestone was exposed and cleared by erosion of overlying beds of dolomite and most glacial drift by erosion. At Fiborn Quarry there is a north-facing 40-foot hillside of rock that is mostly buried in glacial outwash sand, except near the quarry. A shallow rock-floored canyon descends from the quarry area to the lower meandering river channel to the north. This elevation drop was an important factor in facilitating extraction of limestone at Fiborn.

After final retreat of the last ice-age glaciers, and lowering of the glacial (Wisconsin) lake that covered the area, two river channels developed, the South and West branches of Hendrie River. These streams were fed by an extensive wetland ponded against the south side of the exposed Fiborn Limestone, which also appears to have flooded at times over the area of the future quarry, washing off much of the soil. The acidic swamp water, which is able to dissolve the calcite of which limestone is composed, enlarged joints (cracks) in the rock and, over time, produced a network of underground drainage channels, a natural "dewatering" karst drainage system of sinkholes and caves. When the quarry was opened, water flowing into the quarry from the swamp was diverted to these sinkholes, were it was conducted underground to a spring along the South Branch of the Hendrie River. Five active natural "sinks" still performing this drainage function, two fed by creeks named Sinking Creek and Canyon Creek in the quarry, Flat Creek (dug by quarry workers), and Bog and Bridge Creeks west of the quarry. There are a number of other sinkholes, now dry, that probably were once active at higher water levels in the swamp.

An important consideration in the choice of the thickness of limestone to quarry was the concentration of magnesium in the rock. This is expressed as the mass percent of magnesium oxide (MgO) obtained from limestone burned (calcined) at high temperature in a kiln. A pure limestone (100% $CaCO_3$) yields 56.0% CaO (after CO_2 has been driven off by the calcining). A pure dolomite yields 30.4% CaO and 21.9% MgO. The mean composition of the Fiborn Limestone unit expressed in this way is 55.1% CaO and 0.8% MgO. The limestone unit just below the Fiborn Limestone yields 53.9% CaO and 1.4% MgO, which is too much MgO for the manufacture of calcium carbide and was left as the quarry floor.

The Calcium Carbide Connection

Limestone has been used for centuries as a building stone (many municipal buildings are faced with limestone or limestone naturally recrystallized (metamorphosed) to marble, and to make lime plaster. Later, lime from limestone was used in steelmaking and sugar refining. However a discovery in 1892 made Fiborn Limestone particularly valuable. In that year, in Spray, North Carolina, a Canadian-born inventor named Thomas Wilson was trying to produce aluminum metal

(Al) by reducing aluminum oxide (Al_2O_3), with coke (carbon, C), using an electric arc furnace that could produce temperatures of 2,000° F. The practical use of an arc furnace had been made possible only a decade before, in 1882, by Edison's invention of electric dynamos/generators to produce electricity at high power. Wilson found that aluminum was not produced, so he decided to use his furnace to make calcium metal (Ca), by reducing lime (CaO) with coke, which he thought might reduce aluminum oxide to produce aluminum metal. He obtained a molten mass that hardened to a dark gray solid. It wasn't calcium and he threw it away, but noticed when it got wet a highly flammable gas bubbled off (the story is that a workman relieved himself on a pile of this material, while smoking a cigar). A chemist friend of Wilson's identified the gas as acetylene (HCCH), produced by the reaction of water with what Wilson had made, calcium carbide (CaC_2, called hereafter just "carbide"). Wilson obtained patents for the process and built the first factory to produce carbide, using hydroelectric power, in 1895. This accidental invention quickly became a world-wide industry. Acetylene could be burned to produce a very bright flame and was adapted to uses including miners' "carbide lamps" (replacing the candles or oil that had been used since antiquity); bicycle, auto, and train headlamps; work-site illumination; and home lighting (though electric generators quickly replaced the home lighting use of acetylene with cleaner and flame-free electric lights); and in chemical processes. The boom was on: anywhere there was a river with enough flow to produce a large amount of hydroelectric power was considered for carbide production, provided that there was also available limestone of sufficient purity. One such location was at on the Saint Marys River at Sault Ste. Marie.

There had already been an effort to develop a hydroelectric power plant in Sault Ste. Marie, Ontario, and a small one was in operation in 1894 by the Sault Ste. Marie Water and Light Company. That is the same year that Francis H. Clergue arrived with financial backers from the east (where he had developed a successful electric street railway in Bangor, Maine). They bought the hydroelectric plant from the city and then sought to develop other industries to use the power. The main business plan was to build a paper pulp plant to take advantage of the local forests. It was in the course of their business development efforts that news of the invention of a process to make carbide came to their attention: they had electric power and limestone was not too far away across the river. In 1896 the Peoples Gas Light & Coke Company built a carbide plant on the American side of the river and strung cables across it to bring electric power from Ontario's Sault Ste. Marie Water and Light Company.

In 1898 the Union Carbide Company was founded in Virginia to manufacture carbide. In that same year Chase Osborn purchased the Fiborn property and Francis Clergue and associates established the Michigan Lake Superior Power Company (MLSPC) on the American side to build a new and larger hydroelectric plant, with the Union Carbide Company agreeing to lease the second floor of the plant to manufacture calcium carbide (ASME, 1981). The power plant (which still operates) began operations in 1902 with Union Carbide as its only paying customer.

To make a long story short, financial and labor difficulties ensued and in 1903 MLSPC went into bankruptcy. Finally, in 1914, the Union Carbide Company bought out MLSPC and formed the Michigan Northern Power Company. Clearly, carbide manufacture was more lucrative than providing power to other industries that were slow to develop. At one point, the Union Carbide's plant in the "Soo" was the largest producer of carbide in America, with Fiborn Quarry (and other nearby quarries) providing the needed limestone.

It is astonishing how quickly the manufacture of carbide developed after Wilson's discovery. A book about the industry was published in England in 1898 (Thomson, 1898), including advertisements for various devices for producing acetylene. Total world production of carbide reached a maximum of about 8 million tons per year in the 1960s. China is currently the largest producer of carbide for acetylene. A more complete history of the development of the carbide industry is at (ACS, 1998).

Union Carbide operated the Sault plant until 1963, by which time demand for carbide in America had fallen due to competition with other processes for producing acetylene and rapidly decreasing use of carbide in lighting applications. The Sault plant also became less efficient than plants built elsewhere, in particular at Niagara Falls. In that year Union Carbide sold the Sault plant to Edison Sault Electric Company, and left town.

Of course, by that time, Fiborn Quarry had long since stopped providing limestone for making carbide, due in large part to cheaper sources of purer limestone, such as from the famous quarry at Rogers City, Michigan. Today's Fiborn ghost quarry is off the beaten path but it holds a memory of contributing to America's industrial history of innovation, and just as importantly provides a site for understanding the geologic history of karst features in Michigan.

Acknowledgement

I am grateful to geology professor William Neal (emeritus, Grand Valley State University), and Tyrone Black, Michigan DNR geologist, for review comments that improved the content and presentation of this chapter. However I take full responsibility for any omissions or errors that remain.

References

ACS (1998), American Chemical Society National Historic Chemical Landmarks. Discovery of the Commercial Processes for Making Calcium Carbide and Acetylene. (Online at http://portal.acs.org/portal/PublicWebSite/education/whatischemistry/landmarks/calciumcarbideacetylene/index.htm)

ASME (1981), Michigan-Lake Superior Hydro-Power Plant, Sault Ste. Marie, Michigan, American Society of Mechanical Engineers, Brochure H061. (Online at https://www.asme.org/getmedia/9a6fbefb-8d74-4a9d-aaec-f5838421d7e4/61-Michigan-Lake-Superior-Plant.aspx/index.htm)

Catacosinos, P.A. *et al.* (2000), Stratigraphic Nomenclature for Michigan, Michigan Dept. of Environmental Quality, Geological Survey Division and Michigan Basin Geological Society, Lansing, Michigan. (Online at http://www.michigan.gov/documents/deq/2000CHRT_301468_7.PDF)

Dorr, J.A. and D.F. Eschman (1972), Geology of Michigan, University of Michigan Press. 476 p.

Ehlers, G.M. and R.V. Kesling, (1957), Silurian Rocks of the Northern Peninsula of Michigan: Mich. Geol. Soc., Guidebook for Annual Geological Excursion (June). (Online at http://www.mbgs.org/historicalpubsCD_4.html)

Ehlers, G.M. and R.V. Kesling, (1962), Silurian Rocks of Michigan and their Correlation, in Silurian Rocks of the Michigan Basin (1962), Mich. Geol. Soc. Ann. Field Excursion. (Online at http://deepblue.lib.umich.edu/handle/2027.42/48577)

Shaver, R.H. (1996), Silurian Sequence Stratigraphy in the North American Craton, Great Lakes Area: Geological Society of America Special Paper 306. (Online at http://specialpapers.gsapubs.org/content/306/193.full.pdf+html?sid=31d176d2-5265-429d-8777-3665cb090297)

Thompson, G.F. (1898), Acetylene Gas, Nature, Properties and Uses; also Calcium Carbide, its Composition, Properties and Method of Manufacture. (Liverpool, Eng.) (Online at https://play.google.com/books/reader?id=FTBIAAAAIAAJ&printsec=frontcover&output=reader&authuser=0&hl=en&pg=GBS.PR9-IA1)

Endnotes

1   *Michigan: A History of the Wolverine State*, Willis F. Dunbar and George S. May, Wm. B. Eerdmans Publishing 1995 (third edition, first edition 1965), ISBN 0-8028-7055-4.

2   Dunbar and May *Op. Cit.*

3   *The Duluth, South Shore & Atlantic Railway: A History of the Lake Superior District's Pioneer Iron Ore Hauler, John Gaertner,* Indiana University Press 2008(?), ISBN 978-0-253-35192-0. (ISBN-10: 0253351928; ISBN-13: 978-0253351920)

4   Osborn lived at 718 Cedar St. with his family and one servant, according to the 1900 census. He and his wife, Lillian, had five children: Ethel, George, Chase Jr., Emily, and Gertrude.

5   Fitch, who was born in 1839 in Circleville, Ohio, and moved with his parents to Madison, Wisconsin, in 1851, began his railroad career in 1871 as a clerk for the Chicago and Northwestern line. He was appointed general manager of DSS&A by Canadian Pacific President W.C. Van Horne in November, 1888, after CP's takeover of the South Shore, which began in April and was completed in August (Gaertner). Fitch lived with his wife, Emily, daughter Mary, son-in-law Peter W. Phelps, and granddaughter, Emily, along with three servants on Ridge Street in Marquette's third ward, according to the census rolls of 1900 and 1910. Emily Fitch, who died in 1902, was in failing health possibly by the time Fitch met Osborn. The latter's diaries mention several visits by Fitch to Osborn's home over those years, always alone.

6   In August, 1896, Osborn requested and Fitch arranged a special DSS&A sleeper car to take Osborn and his entourage of Sault Ste. Marie Republicans to the party's state convention in Ironwood that month. Osborn sent Fitch a silver toothpick at Christmas, presumably in thanks. They exchanged letters in 1897 about a few matters of railroads and politics, and that November, Fitch authorized a special train to take a doctor to Eckerman, a few miles north of Deerfoot, after an accident at the lodge. Osborn wrote to Fitch in January 1898, noting "No bill has been sent [to] me for the service." Fitch replied "Sorry I cannot accommodate you," – he wasn't going to

send a bill – "but the dates and facts have all slipped my mind and I could not remember them now, if I tried." Expressing "admiration and affection," Osborn had a case of Pommery Sec champagne sent to Fitch from a dealer in Milwaukee.

7　After coming out for the re-election of Gov. Hazen Pingree, Osborn reassured Fitch in a letter that his support for the former Detroit mayor, whose largely frustrated reformist agenda included increasing property taxes on railroads (Dunbar and May), didn't mean he was taking any anti-railroad positions. Even in 1902, when Osborn came out publicly for state ownership of railroads, his correspondence and dealings with Fitch reflected no strain over railroad politics.

8　The August 6, 1898, issue of the journal *The Electrical World* reported Union Carbide's purchase of Lake Superior Carbide Co. and its plan to build a calcium carbide manufacturing plant in Sault Ste. Marie, Michigan.

9　In *The Iron Hunter* Osborn claimed to have no money for a campaign. His diary for the month or so before the convention in Grand Rapids makes no mention of campaigning. Attorney J.E.Whelan nominated Osborn at the convention on Thursday, June 27. The following evening, Osborn took a train from Lansing to Owosso, not Grand Rapids, according to his diary, and spoke to a small group of party officials. "Left 6:20 for Owosso to make Republican speech," he wrote. "Made short talk." Osborn returned to Lansing on Saturday morning (and spent the afternoon watching Albion College win a football game).

10　Dunbar and May. *Op. Cit.* In his autobiography, *The Iron Hunter*, Osborn claimed delegates were being offered $3,000 "when the thing got hot."

11　*Steel at the Sault: Sir James Dunn and the Algoma Steel Corporation 1901-1956*, Duncan L. McDowall, University of Toronto Press 1984, ISBN: 978-080205652-8. (ISBN-10: 0802056520; ISBN-13: 978-0802056528)

12　"This notice was read to John Coward, foreman, and Herbert Sear, laborer at the limestone quarry, on November 16th, 1904, by Nelson Cadarette," a DSS&A worker told Fitch in a telegram. "They stopped work and left for the Soo that night. The only work they had done was to cut down about a dozen trees, clearing a space about 50 feet square."

13　A St. Ignace *Republican-News* story in 1913 said Martin had been actively connected with the stone business for 25 years in Ohio and Kentucky.

14 Osborn travelled throughout the Asian-Pacific region in February, March, and April. His 1905 diary showed him to be in the Philippines and China about the time Fitch was dealing with Martin. During his foreign travels, Osborn frequently wrote dispatches published in papers around the state, starting usually in Sault Ste. Marie, describing faraway lands, natural wonders, and exotic peoples. This trip was no different, and included analyses of the Russo-Japanese war, in which he favored the Russians.

15 In a letter dated August 26, Sault Savings Bank attorney and counselor W.A Coutts asked Osborn about the details of the contract with Martin. "I am entirely satisfied that Martin is liable personally," Coutts wrote, "but whether his relations with you and Mr. Fitch create a liability against you and Mr. Fitch is a question that I cannot determine without further information." Coutts' letter indicated the worker lost all the fingers of one hand.

16 Dunbar and May *Op. Cit.* (1)

17 Gaertner

18 McDowall *Op. Cit.* (11)

19 McDowall *Op. Cit.*

20 Balance sheets for end of June, 1918–1926, are held in the Canadian business history Collection held by the C.B. "Bud" Johnston Library at the University of Western Ontario.

21 *The Michigan Manufacturer* listed the company under "New Michigan Corporations" in its October issue.

22 McDowall *Op. Cit.*

23 Records of the U.S. Bureau of Economic Research.

24 McDowall *Op. Cit.*

25 McDowall *Op. Cit.*

26 The mention in a history of the nearby Wilwin settlement noted that Frank Chesbrough, who bought large tracts of land near Fiborn and helped found Wilwin, investigated the area's limestone resources and was interested in quarrying along with his timber interests, due to the war-time demand. (Chesbrough launched a successful lumber operation, but never built the quarry.) See *The Story of Wilwin*, by William Chesbrough, published online at http://chippewa. migenweb.net/wilwin.htm

27 Dunbar and May *Op. Cit.*

28 Five Fiborn residents' names were drawn: Valentine Hemm, Jack Melnick, Lorence D. Linck, Owen McGraw and someone listed as Bozo Butknich. A few weeks later, the *Republican-News* reported that of those called in the first wave, only six "failed to appear or send

excuse" (for physical examinations), including Butknich, whose first name was now spelled "Buzo." Whoever duped the draft board into calling Bozo/Buzo Butknich to national service apparently was never discovered.

29 The August 11 *Republican-News* listed Walter F. Briere, John Wolminski, Michael Stitchen, Mike Smith, Henry Pope, Leo Franks, Angelo Perone, Thomas Tuomala, and William Richards.

30 About two weeks after reporting that Perone was named in the second call, the *Republican-News'* Fiborn column of September 1 announced "Angelo Peruno has accepted a position with the South Shore in Marquette. Sorry to see you leave, Angelo." (The newspaper's correspondents seemed to have trouble with Angelo's name; the item about draftees spelled his last name "Prueno.") An engineer on Fiborn's locomotives, he was one of hundreds of men to receive a questionnaire from the county draft board late in 1917, so it's possible he was drafted in the wave of men called up based on skills. One of two headstones at his grave in Maplewood Cemetery near Rexton says he served in the 305th Field Artillery, 77th Division. (He died in 1938.) The 305th served in France, and presumably Perone along with it. He returned to Fiborn by 1920, was listed on the census as 27 years old, single, and living at the boarding house. He remained at the quarry at least until the 1930s. According to the census rolls of 1920 and 1930, he could speak English but Italian was his first language.

31 "Following are the contributors: Roy D. Newell, O.P. Welch, Henry Pope, Mike Stitchen, E.B. Judkins, Owen McGraw, George Collins, Wm. Richards, John Elm, Angelo Perrone, Pete Ladane, Fred Cormitch, Fred Kaubanick, Mike Smith, Nick Boll, Mr. and Mrs. Oscar Erickson, Pete Woloski, John Soarmanick, John Walters, John Taylor, D.E. Mason, Oscar Franks, Pat Brady, Edward Cardinal. ... A branch chapter was also organized with the following membership: Roy D. Newell, O.P. Welch, Leo Franks, I.H. Bell, G.A. Cavanaugh, Albert Hunter, Henry Pope, Mike Stitchen, E.B. Judkins, O. McGraw, Geo. Collins, Wm. Richards, John Elm, Angelo Perrone, Fred Kaubanick, Mike Smith, Nick Boll, Oliver Beckland, Mr. and Mrs. O. Erickson, M.J. Campbell, W.A. Prieur, Peter Adatte, David Davidson, Will Nicholson, John Walters, Jack Traynor, D.E. Mason, Oscar Franks, Hugh Miller, Ole Olson, Pete Olson, Pat Brady, Wm. LaFountain, Ed. Cardinal West Buch, (sic) Edw. Franks, John Hill, Herbert Hamilton, Edw. Amyotte." Other quarry employees during the World War I years, gleaned from mentions in the *Republican-News'* Fiborn column: William Webster, John Hill, Daniel Weaver,

and a Mr. and Mrs. Sample, who along with their daughter, Clara, were cooking at "the Fiborn hotel."

32 McDowall *Op. Cit.*

33 Canadian Business History Collection, University of Western Ontario.

34 As well as the Dunham House, Welch became a partner in the city's Hotel Northern along with Hemm. Welch eventually became sole owner of the Northern and one of St. Ignace's most prominent innkeepers, and was an active member of the city's business community through the 1920s and early 1930s. Echoing his support of World War I fundraising drives, Welch was noted in local newspaper accounts as supporting the city's baseball team and being active in civic groups. He retired from the hotel business in 1933 and remained in St. Ignace, often traveling to Ohio or the South in winter, until his death in February, 1937.

35 Algoma Steel reported small, infrequent orders in 1922 and 1923, with its rail mill running only 194 days in 1922, and its rolling mills completely idle. Lake Superior Corp. went from reporting a profit of nearly $6 million in 1921 to a loss of $1.3 million in 1923. The losses continued, again hitting $1.3 million in 1925 (McDowall).

36 "Baseball Almanac" online. The Yankees debuted in their new stadium April 18, 1923. George Herman "Babe" Ruth hit a three-run homer to beat the Boston Red Sox 4–1.

37 Hemm turned up regularly in local newspaper accounts as a board member of a country club, city baseball team manager, and member of the board of public works during the mid-1920s. He and Welch bought a hotel in Sault Ste. Marie and another in St. Ignace, before Hemm sold his interest to Welch and left for Piqua, Ohio, in February, 1927, returning a month later and soon buying an automobile dealership and full-service garage. He ran Hemm Motor Sales and Service until the early 1930s, when he married and moved back to Ohio, visiting St. Ignace and maintaining friendships there for many years.

38 Marcella died in August, 1929, at the age of 31 after an illness that wasn't specified by the St. Ignace *Republican-News*, which reported she underwent surgery at McRae Hospital in Alpena.

39 Other key employees in the mid-late 1920s included: foreman Frank Piaskowski; engineer and later foreman Richard McCullough; and locomotive engineers Fred Welch and Frank Les. Laborers in the late 1920s-early 1930s included: dynamite handler Fred Polinski; night watchman/boiler fireman Fred Huntley; and drill operators

Earl Bigelow (Frank's brother) and Melvin Bigelow (Frank and Earl's father); Albert Ryerse; Wes Bush; Edward Henderson and Alfred Peters. At times during that period, laborers included Fred Moon, Harry Laymon, Joe Bruseau, George Ryerse, Jack Burgess, George Burgess, Fred Pawl, Harry Winters, Irvin Verts, Fred Kolb and Victor Salo (who later lived for many years in Ozark).

40 Don and Vera were natives of Gould City, 40–50 miles southwest of Fiborn, which was founded in 1886 with a lumber mill and grew to a town of 200 by the turn of the century. Don's father, Archibald, owned a lumber operation. Population peaked at 300 in the 1920s, then tumbled as timber ran out and the Depression smothered American industry. Don graduated Ferris Institute in 1910 with a business administration degree. He worked for the Pere Marquette railroad in Detroit before taking a job with Algoma Steel Corp., which included eventual transfer to Fiborn, not far from where he and Vera grew up. Vera was the daughter of Gould City's first postmaster, Adolph Highstone, and was related to the Highstones of St. Ignace, who were prominent in the town's business and political circles.

41 McDowall *Op. Cit.*

42 Canadian Business History Collection, University of Western Ontario.

43 Dunbar and May *Op. Cit.*

44 The plight of Mackinac County's poor was illustrated in a 1934 front-page article in the recently merged *Republican News and St. Ignace Enterprise*, headlined: "Welfare situation has reached crisis." The article described the county's case load rising from 537 families in September to 800 in October, a total of 14,000 people for whom the county had $14,000 that month (down from $14,500 in September). County officials considered two approaches: "One is to cut the wage scale on welfare relief work (work performed by aid recipients) from 50 cents to 35 cents per hour; the other is to halt all work relief projects and put Mackinac's needy on direct relief, which, incidentally would be cheaper to do, but is not accepted generally as a worthy condition to promote." A compromise emerged which placed small families on direct relief for about $20 per month, with larger families getting that plus up to 24 hours work relief, totaling about $33.60 per month.

45 McDowall *Op. Cit.*

46 McDowall *Op. Cit.*

47 McDowall *Op. Cit.*

48 Fiborn Limestone Co. balance sheets in the Canadian Business History Collection at the University of Western Ontario list horses and stable equipment among the company's assets in 1918, but not in 1919 or later.

49 Rose Macco arrived in the U.S. from Italy in 1905 and was listed as "Rosa Mackey" in the 1920 census, a 28-year-old boarder in one of the block homes with quarry foreman Frank Piaskowski, his wife, and mother-in-law. Like Angelo Perone, Macco could speak English but Italian was his first language. Though not listed in the 1930 Fiborn census with Perone, Macco apparently remained in the area. He and Perone are buried next to each other in Maplewood Cemetery near Rexton. Perone died in 1938 at the age of 53, and Macco in 1958 at the age of 69. Clyde Ball recalled Macco as the engineer for whom Ball served as "brakey" on the dinkies.

50 A November 21, 1929, form authorizing the money spent on the cottages showed the cottages cost $740, $770, $1,200 and $1,265, respectively, but the frames for each cost only $25. Rebuilding charges ranged from $550 on the cheaper cottages to $960 and $1,050 for the more expensive ones. Along with painting and electrical wiring, the bill for all four cottages came to $3,980.

51 The *Republican-News* Fiborn column in August, 1917, mentioned a contracting firm building the store, along with the cottages built then.

52 Hendricks Township school census figures for Fiborn:

1907–1911	
1908	5
1909	6
1910	(missing)
1911	(missing)
1912	18
1913	21
1914	20
1915	14
1916	25
1917	(missing)
1918	22
1919	17
1920	15
1921	18
1922	(missing)
1923	(missing)

1924 20
1925 (missing)

53 A list of teachers, gleaned from newspaper accounts, which is likely not complete:

Fidele McLeod *	1910–1912
Florence Cheeseman	Fall 1912
Ellen Leveille	Winter 1913
Fidele McLeod	Fall 1913–1915
Eunice McLeod*	1916 or 1917–1918 or 1919
Bernice Larocque	1919–1920
Miss Leonard	1920–1921
Esther Gustafson	1922–1923
Vera McEachern	1924–1925
Ellen Leveille	1925–1926
Helen Lardie	1926–1927
Elta (Norton)LaCount	1927–1930
Emma Kalnbach	1931–1935

*Fidele and Eunice were sisters.

54 Don and Vera McEachern moved to St. Ignace, then to California, then returned to St. Ignace and ran a resort motel for years. Don also sold insurance. He died in 1965. Vera remained active in civic affairs and was organist at St. Ignace's First Presbyterian Church for nearly 30 years, right up until her death in 1978, at age 85, during a visit to Yosemite in California.

55 1965 interview of Lynn and Lillian Brockway in *The Evening News*. Lynn also ran a sawmill and lumbering operation in Trout Lake, worked for a while at Drummond Dolomite and ran a hardware store in St. Ignace before retiring.

Index